LATERAL FORCES

Robert Marks, P.E., David Eckmann & Stephanie Hautzinger, Contributing Editors

President: Roy Lipner
Vice-President of Product Development and Publishing: Evan M. Butterfield
Editorial Project Manager: Jason Mitchell
Production Coordinator: Daniel Frey
Quality Assurance Editor: David Shaw
Creative Director: Lucy Jenkins

Published by Kaplan AEC Architecture
a division of Dearborn Financial Publishing, Inc.®
A Kaplan Professional Company
30 South Wacker Drive, Suite 2500
Chicago, IL 60606-7481
(312) 836-4400
http://www.ALSonline.com

Printed in the United States of America.

05 06 07 10 9 8 7 6 5 4 3

CONTENTS

INTRODUCTION

Welcome to Kaplan Architecture! Kaplan recently acquired Architectural License Seminars (ALS), the oldest and most respected provider of Architect Registration Examination (ARE) study products. Over 50 percent of the registered architects in practice have used ALS material to prepare for their exam. Kaplan, Inc., is the nation's leading provider of lifelong education, having served more than 30 million individuals over the past 60 years.

THE ARE

Kaplan AEC Architecture provides the only complete, centralized source for all nine divisions of the ARE. All 50 states, five territories, and participating Canadian provinces offer the uniform NCARB Architect Registration Examination (ARE). This exam consists of nine divisions:

- Pre-Design
- General Structures
- Lateral Forces
- Mechanical & Electrical Systems
- Building Design/Materials & Methods
- Construction Documents & Services
- Site Planning
- Building Planning
- Building Technology

The Site Planning, Building Planning, and Building Technology divisions require graphic solutions to vignette graphic problems, while the other six divisions consist of multiple-choice questions. The exams are all administered by computer. Candidates must pass all divisions of the ARE in order to become a registered architect. Those who do not pass a division of the exam may retake it after six months. For further details on the ARE, please visit the NCARB Web site at www.ncarb.org.

Kaplan AEC Architecture provides a variety of study material to help you prepare for the exam: study guides, computer mock exams, paper mock exams, question & answer handbooks, video workshops, and flash cards. All of our products can be ordered online at www.ALSonline.com, or by calling (800) 420-1429.

The ARE is not an easy exam! Although we cannot guarantee a passing grade, we **can** guarantee our material will better prepare you for the ARE.

Good luck on your examination, and in your professional career.

COURSE PURPOSE AND PROCEDURE

This study guide is a condensed and simplified review of Lateral Forces. Courses covering the other divisions of the ARE are also available, all of which have been prepared by the professional staff of ALS.

It should be noted that an understanding of general structures is necessary to properly understand lateral forces.

We advise that you set aside a definite amount of study time each week, just as if you were taking a lecture course, and carefully read all of the material. At the end of each lesson you will find a short review quiz. The quiz questions are intended to be straightforward and objective, and the answers and explanations are included to permit self-checking.

Following the second lesson, there is a Final Examination, which has been designed to simulate the style of the actual exam. Complete answers and explanations will be found on the pages following the examination, to permit self-grading.

OVERVIEW

Structure is, and always has been, an essential part of architecture. From the earliest primitive shelters to today's sophisticated building projects, the goals of architecture have been to satisfy man's inherent sense of beauty, to create spaces appropriate to their function, and—not least of all—to use materials in ways that ensure structural stability against the natural forces of gravity, wind, and earthquake.

During most of human history, people built intuitively, since the physical principles that underlie structural behavior had not yet been propounded. In the past 200 years, however, the science of structures has evolved, so that the design of today's buildings, at best, combines natural intuition with the rigorous mathematics of structural engineering.

It is unfortunate that the study of structures, important as it is, is less than fascinating to most architects. Why this is true is not entirely clear; perhaps the architectural schools emphasize rote problem solving rather than broad concepts. In any event, knowledge of structures by the architect is vital, not only as a necessary step towards the goal of an architectural license, but in his or her professional practice as well. We hope that this course will be informative, perhaps interesting, and that it will be helpful in your preparation for the Architect Registration Examination.

This edition of *Lateral Forces* incorporates references to both the 1994 and 1997 editions of the Uniform Building Code.

States and municipalities across the country have begun adopting the International Code Council's International Building Code. While this code may eventually replace the Uniform Building Code, the ARE only requires you to familiarize yourself with any *one* of the codes in use today. Any question on the ARE involving codes will provide you with the table or equation you need to answer it correctly. For readers interested in comparing codes, this edition of *Lateral Forces* includes an appendix featuring several tables used or referenced by the IBC.

This course was written by Robert Marks, P.E. Mr. Marks is a structural engineer in private practice in Los Angeles and a founding partner of ALS.

You are now ready to start your course in *Lateral Forces.* We wish you success in your examination.

ABBREVIATIONS

The following abbreviations and symbols are used in this course and are generally understood in structural design practice.

Abbreviation	It is read
ft. or '	foot
ft.^{2} or sq.ft.	square foot
ft.^{3} or cu.ft.	cubic foot
ft.-kip or 'k	foot-kip
ft.-lb. or '#	foot-pound
in. or "	inch
in.^{2} or sq.in.	square inch
in.^{3} or cu.in.	cubic inch
in.-kip or "k	inch-kip
in.-lb. or "#	inch-pound
kip or k	kip (1 kip = 1 kilo pound or 1,000 pounds)
ksi	kips per square inch
lb. or #	pound
lb./cu.ft. or #/ft.^{3} or pcf	pound per cubic foot
plf or #/' or #/ft.	pounds per lineal foot
psf or #/ft.^{2}	pounds per square foot
psi or #/in.^{2}	pounds per square inch
yd.^{3} or cu.yd.	cubic yard
Σ	summation of
$>$	is greater than
$<$	is less than

LESSON ONE

EARTHQUAKE DESIGN

INTRODUCTION

The Lateral Forces division of the ARE tests candidates primarily for their knowledge of earthquake and wind design. Whether or not you live in an area susceptible to earthquakes, you must have sufficient understanding of earthquake design to pass this test. The purpose of this lesson is to provide you with that understanding. The equally important subject of wind design is covered in the next lesson.

The following are reproduced in this lesson from the 1994 edition of the Uniform Building Code by permission of the International Conference of Building Officials: Tables 16-I, 16-J, 16-K, 16-L, 16-M, 16-N, 16-O, 23-I-J-1, and 23-I-K-1, and Figures 16-2 and 16-3.

NATURE OF EARTHQUAKES

Of all the natural disasters that can occur, earthquakes are probably the most feared and the least predictable. With little or no warning, tremendous forces are unleashed that last only seconds, yet can cause widespread destruction of life and property.

Although advances have been made in the prediction of earthquakes, they remain capricious: reliable earthquake prediction is still a long way off. And we still have no way of controlling the vast amount of energy released by an earthquake. However, the detailed study of many earthquakes in recent years by seismologists and engineers has brought about an increased understanding of the physical effects of earthquakes on buildings and other structures. As a result, our ability to design and construct buildings to resist earthquakes has significantly improved.

What causes earthquakes? Centuries ago, people believed that these inexplicable movements of the earth were caused by the restlessness of a monster that supported the world on its back. Depending on your tribe, the monster might be an immense frog, a giant spider, or a huge tortoise. Until comparatively recent times, seismology, the study of earthquakes, had not advanced much beyond those primitive beliefs.

Even now, the causes of earthquakes are not totally understood. However, the theory which best explains earthquake phenomena is called *plate tectonics*. According to this theory, the earth's crust is made up of great rock plates floating on a layer of molten rock. As these plates move with respect to each other, friction and other forces lock them together. As the strain increases over long periods of time, the frictional force between the plates is finally overcome and the plates violently slip past each other to a new unstrained position. This slip releases complex shock waves, which travel outward at great speeds through the rock and overlying soil. The ground through which these waves travel experiences shaking or vibration, and buildings founded in this shaking ground are pushed and pulled to-and-fro, side-to-side, and up-and-down. In a short time, perhaps 10 to 20 seconds, the worst part of the earthquake is usually over, often followed by a number of aftershocks of lesser severity. The intent of earthquake, or seismic, design is to provide buildings with sufficient resistance to the effects of this ground shaking so that they can survive even a major earthquake without collapsing.

About 80 percent of the world's earthquakes, including many of the strongest on record, occur around the rim of the Pacific Ocean, from the tip of South America up through Alaska and down along Japan and China to the South Pacific. In the United States, the most vulnerable areas are in California and Alaska. However, many locations east of the Rockies are also subject to high seismic risk: one of the most severe earthquakes in our history occurred in the Mississippi River valley near the confluence of the Ohio and Mississippi Rivers in 1811.

The term *fault* refers to the boundary between adjacent tectonic plates, such as the San Andreas fault in California. It also denotes the numerous lesser breaks and fractures in the earth's crust. A major fault, such as the San Andreas, is frequently a system of faults comprising a fault zone of considerable width. A strong earthquake may result in lateral or vertical shifts along the fault of anywhere from a few inches to as much as 20 feet.

The location in the earth's crust where the rock slippage begins is called the *focus* or *hypocenter* of the earthquake. Its depth is usually five to ten miles in California, although foci several hundred miles deep have been located in other parts of the world. The projection of the focus on the ground surface is called the *epicenter*.

MEASUREMENT OF EARTHQUAKES

Earthquakes are measured by two distinctly different scales: the *Richter scale* and the *Modified Mercalli scale*. The Richter scale is a measure of the earthquake's magnitude—the amount of energy it releases. The scale is logarithmic, with each number representing about 33 times the energy of the lower number. Thus, an earthquake of Richter magnitude 7 releases 33 times more energy than one of magnitude 6, or about 1,000 times (33 × 33) more energy than one of magnitude 5.

The largest earthquakes ever recorded had a magnitude of about 8.9. The 1906 San Francisco earthquake has been estimated at about 8.3, the Alaska earthquake of 1964 measured 8.4, the San Fernando earthquake of 1971 registered 6.4, the Northern California earthquake of 1989 had a magnitude of 7.1, and the 1994 Northridge earthquake measured 6.7.

The Richter scale does not provide an adequate indication of the potential damage to buildings: an earthquake of moderate magnitude centered in a densely populated area would be more damaging than a greater quake located far from population centers.

The damage effects of great earthquakes may not be much greater than those of smaller quakes; the difference is that the greater quakes generally last longer and cause damage over a much greater area.

Earthquakes of magnitude 5 are potentially damaging, 6 to 7 are considered moderate and potentially destructive, 7 to 7.75 are termed *major*, while those above 7.75 are called *great earthquakes*.

The Modified Mercalli scale measures the intensity of an earthquake (its effects on people and buildings). It is expressed in Roman numerals I to XII, each number corresponding to a specific description of the earthquake's effects. For example, I is *not felt except under especially favorable circumstances*, while XII has *damage nearly total. Large rock masses displaced. Lines of sight and level distorted. Objects thrown into the air.*

There is no precise correlation between the Richter and the Modified Mercalli scales. The former is a fairly accurate measure of physical energy since it is determined by instruments, but it does not tell us about the effects of the earthquake.

The latter, on the other hand, gives us an indication of damage, but it is based on personal observation only, and is therefore imprecise.

RESPONSE OF BUILDINGS TO EARTHQUAKES

An earthquake causes the ground to shake erratically, both vertically and horizontally. The vertical motions are usually neglected in design since they are generally smaller than the horizontal. Also, since buildings are basically designed for vertical gravity loads, they usually have considerable excess strength in the vertical direction.

The ground does not move at a uniform velocity. Quite the contrary: its velocity is constantly changing during an earthquake. The rate of change of velocity is called *acceleration*. It is convenient to measure this acceleration as a fraction or percentage of g, the acceleration of gravity. (The acceleration of gravity is the acceleration experienced by a freely falling body and is equal to 32 feet per second per second.) Thus, we may say that the ground acceleration at a given time is .10 g or 10 percent g, which is equal to .10 × 32 = 3.2 feet per second.

Information on ground acceleration during an earthquake is recorded on *strong-motion accelerographs.* These seismological instruments are normally inoperative; when subject to strong earth motion, however, they become activated, record the earth motion, and then shut off. As recorded by these instruments, the ground acceleration during an earthquake changes rapidly and irregularly, varying from zero to a maximum in each direction.

If these varying ground accelerations are applied to a series of idealized structures, each having a different natural period, the maximum acceleration of each of these structures can be determined, and the results plotted.

The resulting curve is called a *response spectrum.* Figure 16-3 of the Uniform Building Code, reproduced above right, shows three response spectra, one for each of three soil types. While the spectrum will be different for every accelerograph record and hence every earthquake, the spectra all have similar shapes, from which certain conclusions may be drawn.

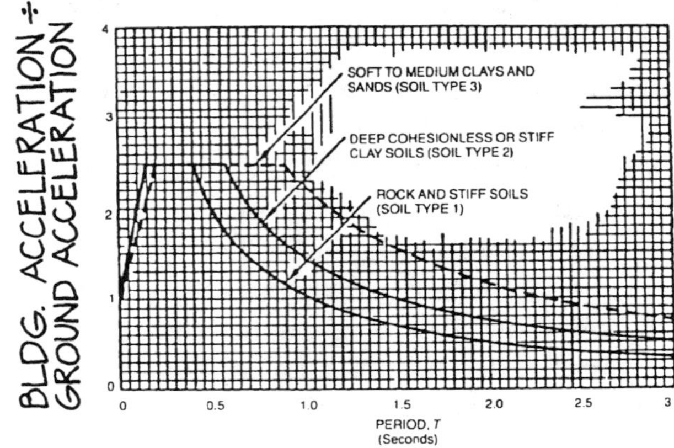

NORMALIZED RESPONSE SPECTRA SHAPES

Reprinted from *Uniform Building Code,* copyright International Conference of Building Officials, Whittier, California, 1994. All rights reserved.

Let us digress for a moment to explain what is meant by *period.* If you take a pendulum, consisting of a weight or mass on top of a rod with a certain stiffness, and you start the pendulum moving, it will take a certain amount of time for it to go through one complete back-and-forth motion. This amount of time is called its *natural period* or *fundamental period* and is independent of how hard you push on the pendulum.

The natural period is dependent on only two factors: the mass, or weight, of the pendulum and the stiffness, or rigidity, of the rod. Thus, making the mass heavier or the rod more flexible changes the natural period of the pendulum.

Although a building is vastly more complex than a simple pendulum, it also has a natural period, measured in seconds, which is dependent only on the mass and stiffness of the building. The relationship between the stiffness and natural period is inverse. In other words, *increasing the stiffness of a building, or making it more rigid, decreases its period.* Conversely, *making the building more flexible, or decreasing its stiffness, increases its natural period.* On the other hand, the period and the mass of the building have a direct relationship; an increase in the building's mass results in an increase in the natural period.

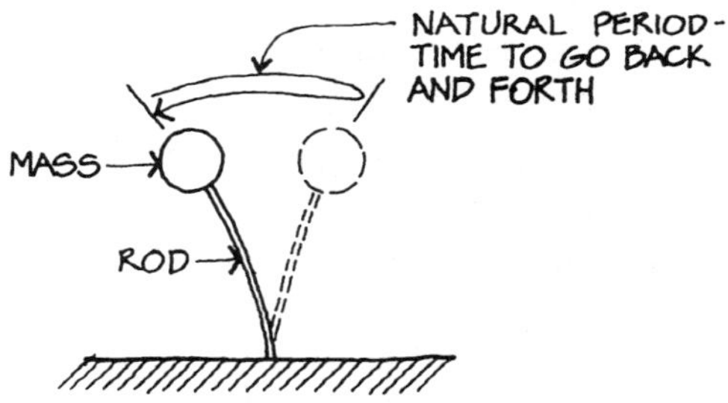

PENDULUM

Getting back to the response spectra, what they tell us is that for a structure having a natural period of 0 seconds (an infinitely rigid structure), the maximum acceleration of the structure will be the same as that of the ground. As the period of the structure increases, its peak acceleration also increases, reaching a maximum when the period is about 0.5 seconds. This maximum acceleration is generally two to three times the maximum ground acceleration. However, even larger magnification of ground acceleration is possible. In the Mexico City earthquake of 1985, for example, accelerations in the upper floors of some buildings were six times the maximum ground acceleration.

As the period increases beyond 0.5 seconds or so, the acceleration decreases and soon becomes less than the ground acceleration. Thus, in general, we can say that a building that is less stiff, or *more flexible*, will have a *longer natural period* and hence a *lower maximum acceleration*.

The response spectra also show that the building acceleration will be greater if the building is on soft ground (soil type 3) and less if it is on rock or stiff soil (soil type 1).

Imagine for a moment an earthquake in which the ground moves instantaneously in a given horizontal direction. A building sitting on the ground will momentarily have its base offset relative to the upper portion of the building. The physical property that causes the upper portion of the building to remain in its original position while the base is moved is called *inertia*.

The same distorted shape of the building could be obtained by applying simulated horizontal forces to the upper portion of the building. The internal forces created by these simulated external forces are equal in magnitude to the actual forces induced in the structure by the earthquake motion. In other words, the applied earthquake forces are not real externally applied forces, as are dead, live, and wind forces, but are simulated forces that produce the same effect on the building structure as the actual earthquake motion.

If we know the building's natural period, we can determine its acceleration for a given earthquake by using the response spectrum.

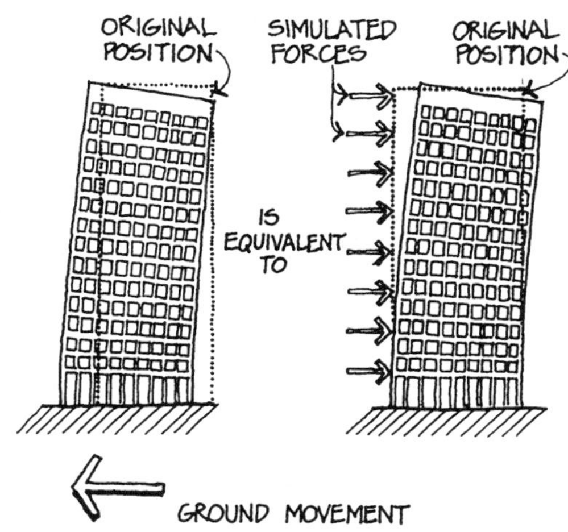

SIMULATED EARTHQUAKE FORCES

We can then convert acceleration to force by using Newton's second law of motion: F = Ma, or force is equal to mass times acceleration. Mass is weight divided by g, the acceleration of gravity (g = 32 feet per second per second). *Thus, the magnitude of the inertia force is equal to the mass of the structure multiplied by its acceleration.* Since long period, or flexible, buildings have less acceleration than short period, or stiff, buildings, they induce less force. And since heavy buildings have more mass than light buildings, they have greater earthquake forces. It can be seen, therefore, that there is a significant difference between conventional structural design and seismic design. In the former, the magnitude and nature of the applied dead, live, and wind loads are *independent* of the stiffness of the structure. In seismic design, however, increasing the structure's stiffness increases the induced seismic force; the seismic force is therefore *dependent* on the stiffness of the structure. If we have some understanding of these principles, we will see the rationale behind the Uniform Building Code earthquake requirements.

CODE REQUIREMENTS

The code most commonly used for earthquake design is the Uniform Building Code (UBC), and the discussion that follows is based on the 1994 edition of this code.

Introduction

Structures designed in accordance with the UBC provisions should generally be able to:

1. Resist minor earthquakes without damage.
2. Resist moderate earthquakes without structural damage, but possibly some nonstructural damage.
3. Resist major earthquakes without collapse, but possibly some structural and nonstructural damage.

The code is intended to safeguard against major failures and loss of life; the protection of property per se is not its purpose. While it is believed that the code provides reasonably for protection of life, even that cannot be completely assured.

The code provides for resistance to earthquake ground shaking only, but not for earth slides, subsidence, settlement, faulting in the immediate vicinity of the structure, or for soil liquefaction, which is the transformation of soil into a liquefied state similar to quicksand.

The UBC provides two methods for determining the earthquake forces: a dynamic lateral force procedure and a static lateral force procedure.

The dynamic procedure is always acceptable for design. The static procedure is allowed only under certain conditions of building regularity, occupancy, and height. For example, regular buildings under 240 feet in height may be designed on this basis.

Collapsed apartment building, San Francisco, 1989

Reproduced with permission. Copyright the National Oceanic and Atmospheric Administration.

In the dynamic lateral force procedure, a mathematical model of the structure is developed and then subjected to appropriate ground motions. A computer program determines the model's dynamic response to the input motions, from which the forces to be used in the design can be determined. The details of the dynamic procedure are quite technical and therefore beyond the scope of this course.

Since the static procedure may be used whenever a dynamic procedure is not required, it is the method used in the seismic design of most buildings. In this method, we apply static or non-varying horizontal forces to the structure that produce internal forces similar to those that would be induced by actual earthquake motion. As discussed previously, during an earthquake there are no external forces actually applied to the building; the forces used in a static analysis only simulate the effects of the earthquake, and are not real forces, as are dead, live, and wind loads.

Static Lateral Force Procedure

The UBC static procedure consists of determining the total lateral force, or shear at the base (V) from the following formula:

$$V = \frac{ZIC}{R_w} W$$

where

V = the total lateral force or shear at the base
Z = the seismic zone factor
I = the importance factor
C = a numerical coefficient
$= \frac{1.25S}{T^{2/3}}$
R_w = a numerical coefficient
S = the site coefficient for soil characteristics
T = the fundamental period of vibration, in seconds, of the structure in a given direction
W = the total dead load

It might help to understand this formula if we go back to the basic relationship between force and acceleration, which is F = Ma. This is equivalent to V = ZCW.

The importance factor I is introduced, in order to increase V for certain critical structures. Thus, V = ZICW.

Now we take that value of V and reduce it by the coefficient R_w, to account for the ability of the structural system to accommodate loads and absorb energy considerably in excess of the usual allowable stresses without collapsing, and thus

$$V = \frac{ZIC}{R_w} W, \quad \text{the basic seismic formula.}$$

In the paragraphs that follow, we will explain the terms in this formula in more detail.

The seismic force V is generally evaluated in the two horizontal directions parallel to the axes of the building. The structure must be adequate to resist the stresses caused by the seismic force in each direction, but not in both directions acting at the same time.

Z Factor

The seismic zone map of the United States is Figure 16-2 in the 1994 UBC, which is reproduced on the following page. You will note that the country is divided into six seismic zones: 0, 1, 2A, 2B, 3, and 4, according to the seismic risk in each zone. Thus, Zone 0 (the Gulf coast) is considered to have no seismic risk, while Zone 4 (primarily in California and Alaska) has the greatest seismic risk. The seismic zone factor Z varies from 0 in Seismic Zone 0 to 0.40 in Seismic Zone 4, as shown in Table 16-I, reproduced on the following page.

The Z factor corresponds numerically to the effective peak ground acceleration (EPA) in each zone, expressed as a fraction of g, the acceleration of gravity. Therefore, the effective peak ground acceleration in Zone 4 = Z in Zone 4 = 0.40 g, or 40 percent of g.

I Factor

I is the importance factor as defined by Table 16-K, reproduced on page 10 (IBC Table 1604.5, see Appendix).

You can see that the value of I for earthquake is either 1.0 or 1.25, depending on the occupancy category of the building.

Hence, essential facilities such as hospitals and fire and police stations are designed for seismic forces 25 percent greater than normal (I = 1.25). In this way, such emergency facilities are expected to be safe and usable following an earthquake.

Similarly, hazardous facilities must also be designed for these increased seismic forces, in

TABLE 16-I—SEISMIC ZONE FACTOR Z

ZONE	1	2A	2B	3	4
Z	0.075	0.15	0.20	0.30	0.40

The zone shall be determined from the seismic zone map in Figure 16-2.

FIGURE 16-2—SEISMIC ZONE MAP OF THE UNITED STATES

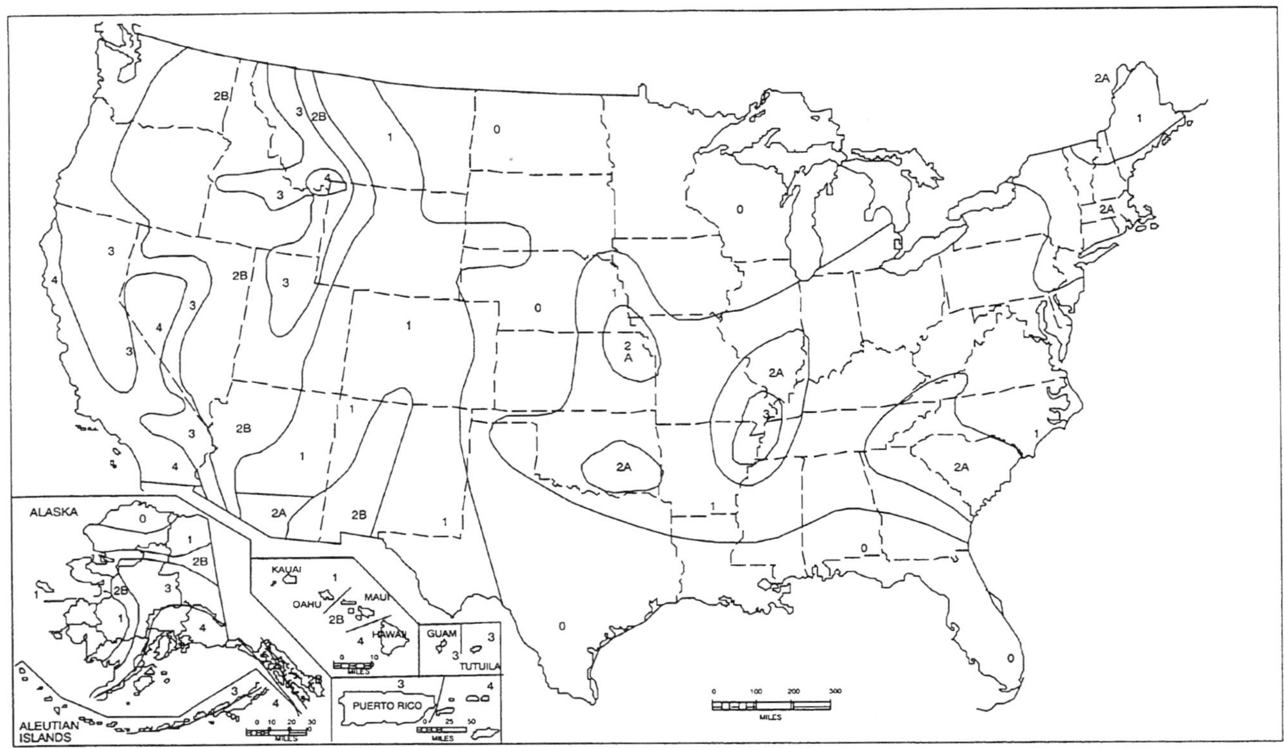

order to provide a greater measure of safety for such facilities.

The importance factor for wind will be discussed in Lesson 2.

C Coefficient

The response spectra on page 4 show that a building with a short period has an acceleration greater than the ground acceleration, while a long period building has an acceleration less than the ground acceleration. Furthermore, buildings on soft soil have greater accelerations than those on rock or stiff soil.

These relationships are quantified by using the numerical coefficient C, which is equal to $\frac{1.25S}{T^{2/3}}$. In other words, C modifies the ground acceleration, and thus the resulting seismic force, to account for the soil characteristics of the site (S) and the building's period (T).

The site coefficient S is determined from Table 16-J, reproduced on the following page. Its value generally varies from 1.0 to 1.5. Soil type

TABLE 16-J—SITE COEFFICIENTS[1]

TYPE	DESCRIPTION	S FACTOR
S_1	A soil profile with either: (a) A rock-like material characterized by a shear-wave velocity greater than 2,500 feet per second (762 m/s) or by other suitable means of classification, or (b) Medium-dense to dense or medium-stiff to stiff soil conditions, where soil depth is less than 200 feet (60 960 mm).	1.0
S_2	A soil profile with predominantly medium-dense to dense or medium-stiff to stiff soil conditions, where the soil depth exceeds 200 feet (60 960 mm).	1.2
S_3	A soil profile containing more than 20 feet (6096 mm) of soft to medium-stiff clay but not more than 40 feet (12 192 mm) of soft clay.	1.5
S_4	A soil profile containing more than 40 feet (12 192 mm) of soft clay characterized by a shear wave velocity less than 500 feet per second (152.4 m/s).	2.0

[1]The site factor shall be established from properly substantiated geotechnical data. In locations where the soil properties are not known in sufficient detail to determine the soil profile type, soil profile S_3 shall be used. Soil profile S_4 need not be assumed unless the building official determines that soil profile S_4 may be present at the site, or in the event that soil profile S_4 is established by geotechnical data.

S_4 (S = 2.0) need not be assumed unless there is evidence that deep deposits of soft clays exist at the site. If the soil properties are not known, soil type S_3 is usually assumed, which has an S value of 1.5.

The value of the fundamental period of vibration of the building (T) may be approximated from the following formula:

$$T = C_t(h_n)^{3/4}$$

where

C_t = 0.035 for steel moment-resisting frames

C_t = 0.030 for reinforced concrete moment-resisting frames and eccentric braced frames

C_t = 0.020 for all other buildings

h_n = height of building

Note that in the formula for T, h_n is raised to the 3/4 power. And in the formula for C, T is raised to the 2/3 power. These calculations must be done on a calculator that has the function of raising numbers to a power.

The maximum value of C is 2.75, and this value may be used for any structure regardless of soil type or building period. Thus, in cases where it is not practical to evaluate the site soil conditions and the building period, the seismic load may easily be determined.

The minimum value of the ratio C/R_w is generally 0.075. This minimum design requirement is intended to assure that long period buildings are designed for adequate seismic forces.

The product ZC represents the acceleration response spectrum which has a 90 percent probability of not being exceeded in 50 years.

R_w Coefficient

When we design a structural system to resist dead, live, and wind loads, we attempt to keep deformations and stresses within acceptably low limits. However, it would be economically pro-

TABLE 16-K—OCCUPANCY CATEGORY

OCCUPANCY CATEGORY	OCCUPANCY OR FUNCTIONS OF STRUCTURE	SEISMIC IMPORTANCE FACTOR, I	SEISMIC IMPORTANCE[1] FACTOR, I_p	WIND IMPORTANCE FACTOR, I_w
1. Essential facilities[2]	Group I, Division 1 Occupancies having surgery and emergency treatment areas Fire and police stations Garages and shelters for emergency vehicles and emergency aircraft Structures and shelters in emergency-preparedness centers Aviation control towers Structures and equipment in government communication centers and other facilities required for emergency response Standby power-generating equipment for Category I facilities Tanks or other structures containing housing or supporting water or other fire-suppression material or equipment required for the protection of Category I, II or III structures	1.25	1.50	1.15
2. Hazardous facilities	Group H, Divisions 1, 2, 6 and 7 Occupancies and structures therein housing or supporting toxic or explosive chemicals or substances Nonbuilding structures housing, supporting or containing quantities of toxic or explosive substances which, if contained within a building, would cause that building to be classified as a Group H, Division 1, 2 or 7 Occupancy	1.25	1.50	1.15
3. Special occupancy structures[3]	Group A, Divisions 1, 2 and 2.1 Occupancies Buildings housing Group E, Divisions 1 and 3 Occupancies with a capacity greater than 300 students Buildings housing Group B Occupancies used for college or adult education with a capacity greater than 500 students Group I, Divisions 1 and 2 Occupancies with 50 or more resident incapacitated patients, but not included in Category I Group I, Division 3 Occupancies All structures with an occupancy greater than 5,000 persons Structures and equipment in power-generating stations; and other public utility facilities not included in Category I or Category II above, and required for continued operation	1.00	1.00	1.00
4. Standard occupancy structures[4]	All structures housing occupancies or having functions not listed in Category I, II or III and Group U Occupancy towers	1.00	1.00	1.00
5. Miscellaneous structures	Group U Occupancies except for towers	1.00	1.00	1.00

[1]The limitation of I_p for panel connections in Section 1631.2.4 shall be 1.0 for the entire connector.
[2]Structural observation requirements are given in Sections 108, 1701 and 1702.
[3]For anchorage of machinery and equipment required for life-safety systems the value of I_p shall be taken as 1.5.

Reprinted from *Uniform Building Code,* copyright International Conference of Building Officials, Whittier, California, 1994.

hibitive to use those same limitations when designing for the maximum expected earthquake motion. Instead, the basic philosophy of seismic design is that the structure be able to accommodate the maximum expected earthquake *without collapse*. Although the structure is expected to ride out the earthquake, inelastic deformation is expected to occur, as well as structural and nonstructural damage.

The numerical coefficient R_w, which is determined by the type of lateral load resisting system used, is a measure of the system's ability to accommodate earthquake loads and absorb energy without collapse. A stiff, brittle structure has a low value of R_w, while a resilient, ductile system has a high value of R_w. Later in this lesson we will discuss the R_w values of various lateral load resisting systems in more detail.

W Factor

W is the total dead load and applicable portions of other loads as follows:

1. In storage and warehouse occupancies, 25% of the floor live load is applicable.
2. Where a partition load is used in the floor design, 10 psf is included.
3. Where the snow load is greater than 30 psf, the snow load is included, subject to reduction under certain conditions.

LATERAL LOAD RESISTING SYSTEMS

Introduction

When earthquake motion causes a building to move, energy is induced into the building structure. The function of the lateral load resisting system is to absorb this energy by moving, or deforming, without collapse. The building will probably be damaged during a major earthquake, but the damage is expected to be repairable.

The ability of structural systems and materials to deform and absorb energy, without failure or collapse, is termed *ductility*. Materials or systems that are able to absorb energy through movement are called *ductile*, whereas those which are less able to do so are termed *non-ductile* or *brittle*. An extreme example of a ductile, though non-structural, material would be a rubber band, while an example of a non-ductile material would be unreinforced concrete.

All structural systems resist forces in three basic ways: by bending (flexure), shear, or axial tension and compression. In general, systems that resist forces by bending are more ductile than those that are stressed in shear or axial tension and compression.

There are three basic types of lateral load resisting systems: moment-resisting frames, shear walls, and braced frames. Generally, shear walls are the most rigid, that is, they deflect the least when subject to a given load. Braced frames are usually less rigid than shear walls, and moment-resisting frames are the least rigid.

Moment-Resisting Frames

A moment-resisting frame resists lateral forces by bending action, as shown on the following page.

Moment-resisting frames may be constructed of either structural steel or reinforced concrete.

There are three categories of moment-resisting frames:

1. A *special moment-resisting frame* (SMRF) is a moment-resisting frame made of structural steel or reinforced concrete that has the ability to absorb a large amount of energy in the inelastic range, that is, when the material is stressed above its yield point, without failure and without deforming unacceptably. Concrete frames in Seismic Zones 3 and 4 must be special moment-resisting frames.
2. An *intermediate moment-resisting frame* (IMRF) is a concrete frame that has less stringent requirements than a SMRF, and in

general may only be used in Seismic Zones 1 and 2.

3. An *ordinary moment-resisting frame* (OMRF) is a steel or concrete frame that does not meet the special detailing require-ments for ductile behavior. It may be used in any seismic zone for steel and in Zone 1 only for concrete.

Special moment-resisting frames of concrete are designed and detailed to assure ductility, and the code requirements for such frames are quite specific, including the use of special transverse reinforcement, the use of top and bottom reinforcement throughout the length of beams, special attention to details such as reinforcement splices, and very close special inspection.

Such frames are usually cast-in-place; however, precast beam-column elements are permitted if it can be shown that they will have the same capabilities as cast-in-place concrete.

A moment-resisting frame made of structural steel does not automatically qualify as a special moment-resisting frame. It must comply with special criteria in the code, including requirements for the design of beam-to-column connections, requirements for nondestructive testing of welded connections, and limiting the material that may be used to certain specified grades of steel only.

Since a moment-resisting frame resists lateral loads by bending, it is the most ductile lateral load resisting system.

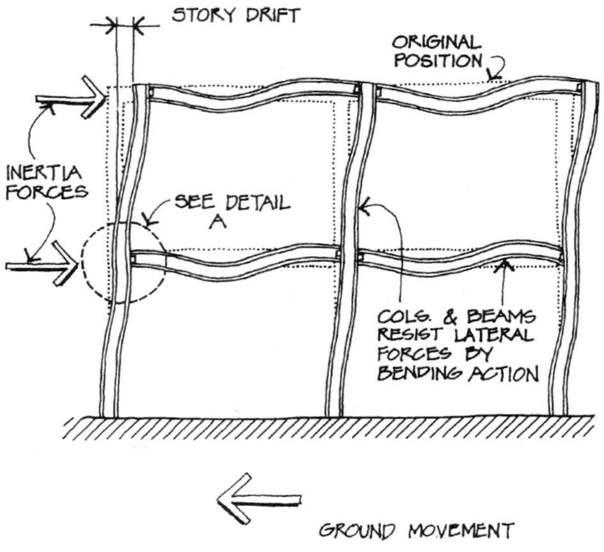

MOMENT-RESISTING FRAME

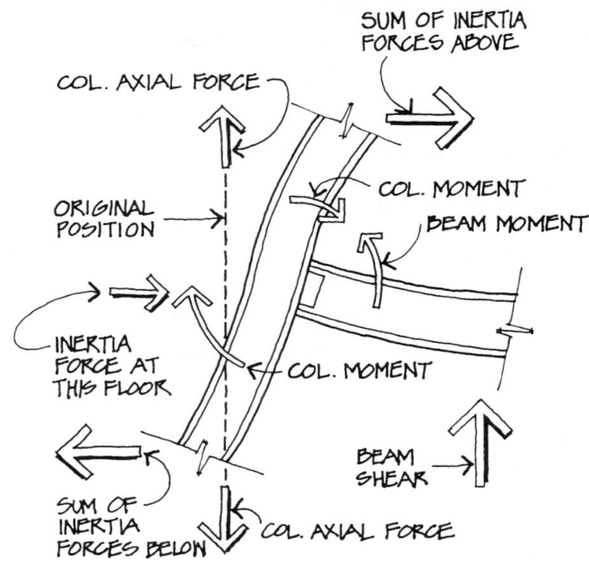

DETAIL A

Shear Walls

A shear wall is one that resists lateral forces by developing shear in its own plane and cantilevering from its base, as shown on the following page.

Thus, a shear wall is essentially a very deep cantilever beam that develops flexural stresses in addition to its basic shear stress. The wall tends to lift up at one end, and there must be sufficient dead load to prevent this, or if not, it must be adequately tied down to its foundation. Also, the connection of the wall to the foundation must be sufficient to prevent sliding.

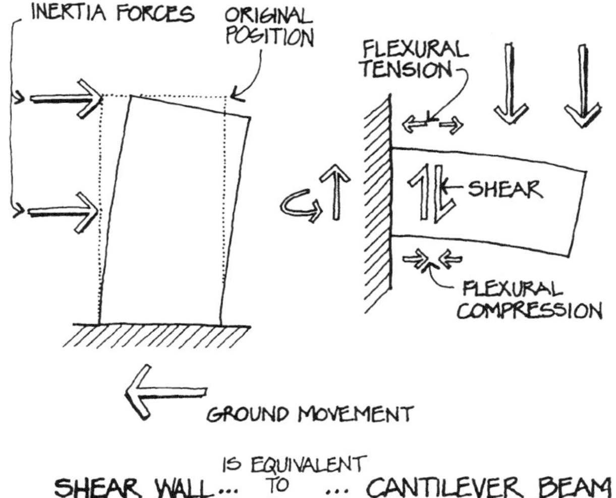

Shear walls may be made of reinforced concrete, reinforced masonry, or steel. They may also be constructed of a facing of plywood, particleboard, or fiberboard applied over wood studs. The allowable shear in a shear wall constructed of plywood sheathing over wood studs is found in Table 23-I-K-1 of the UBC (IBC Table 2306.4.1, see Appendix), reproduced on page 14. In this table, the term *wood structural panel* refers to various wood-based structural panels, including plywood.

For example, the allowable shear in a plywood shear wall using 15/32" plywood of Structural I grade with 8d nails spaced at six inches at plywood panel edges is 280 pounds per foot.

Reinforced concrete and reinforced masonry shear walls generally have allowable shears much greater than those for plywood shear walls.

Braced Frames

A braced frame is a vertical truss that resists lateral forces by axial tension and compression in the truss members. Braced frames are most often constructed of structural steel, although reinforced concrete or timber braced frames are possible.

Collapsed double-deck section of Nimitz Freeway, Oakland, 1989

TABLE 23-I-K-1—ALLOWABLE SHEAR FOR WIND OR SEISMIC FORCES IN POUNDS PER FOOT FOR WOOD STRUCTURAL PANEL SHEAR WALLS WITH FRAMING OF DOUGLAS FIR-LARCH OR SOUTHERN PINE[1,2]

PANEL GRADE	MINIMUM NOMINAL PANEL THICKNESS (inches)	MINIMUM NAIL PENETRATION IN FRAMING (inches)	PANELS APPLIED DIRECTLY TO FRAMING					PANELS APPLIED OVER 1/2-INCH (13 mm) OR 5/8-INCH (16 mm) GYPSUM SHEATHING				
			Nail Size (Common or Galvanized Box)	Nail Spacing at Panel Edges (in.) × 25.4 for mm				Nail Size (Common or Galvanized Box)	Nail Spacing at Panel Edges (in.) × 25.4 for mm			
	× 25.4 for mm			6	4	3	2[3]		6	4	3	2[3]
				× 0.0146 for N/mm					× 0.0146 for N/mm			
Structural I	5/16	1 1/4	6d	200	300	390	510	8d	200	300	390	510
	3/8	1 1/2	8d	230[4]	360[4]	460[4]	610[4]	10d[5]	280	430	550	730
	7/16			255[4]	395[4]	505[4]	670[4]					
	15/32			280	430	550	730					
	15/32	1 5/8	10d[5]	340	510	665	870	—	—	—	—	—
C-D, C-C Sheathing, plywood panel siding and other grades covered in U.B.C. Standard 23-2 or 23-3	5/16	1 1/4	6d	180	270	350	450	8d	180	270	350	450
	3/8			200	300	390	510		200	300	390	510
	3/8	1 1/2	8d	220[4]	320[4]	410[4]	530[4]	10d[5]	260	380	490	640
	7/16			240[4]	350[4]	450[4]	585[4]					
	15/32			260	380	490	640					
	15/32	1 5/8	10d[5]	310	460	600	770	—	—	—	—	—
	19/32			340	510	665	870					
			Nail Size (Galvanized Casing)					Nail Size (Galvanized Casing)				
Plywood panel siding in grades covered in U.B.C. Standard 23-2	5/16	1 1/4	6d	140	210	275	360	8d	140	210	275	360
	3/8	1 1/2	8d	160	240	310	410	10d[5]	160	240	310	410

There are two types of braced frames: *concentric* and *eccentric.*

In the concentric braced frame, which has been widely used for many years, the center lines of all intersecting members meet at a point, and therefore the members are subjected primarily to axial forces.

The eccentric braced frame (EBF) is a steel braced frame in which at least one end of each brace is eccentric to the beam-column joint or the opposing brace. The intent is to make the braced frame more ductile and therefore able to absorb a significant amount of energy without buckling the braces. Examples of concentric and eccentric braced frames are shown on the following page.

Because shear walls and concentric braced frames resist lateral load by shear and axial forces, they cannot undergo as much motion without collapse, or absorb as much energy, as a moment-resisting frame. Hence, these systems are not as ductile as moment-resisting frames.

On the other hand, eccentric braced frames, if properly designed and detailed, can perform in a ductile manner similar to moment-resisting frames.

R_w Values

Now that we have described the different types of lateral load resisting systems, we can refer to Table 16-N of the UBC (IBC Table 1617.6.2, see Appendix), reproduced on page 16, to determine the R_W values for different systems. First we need to define a few terms.

A *bearing wall system* is one in which the vertical load is supported by bearing walls or bracing systems and the required lateral forces are resisted by shear walls or braced frames. In other words, the shear walls or braced frames support vertical, as well as lateral, loads. If these elements fail

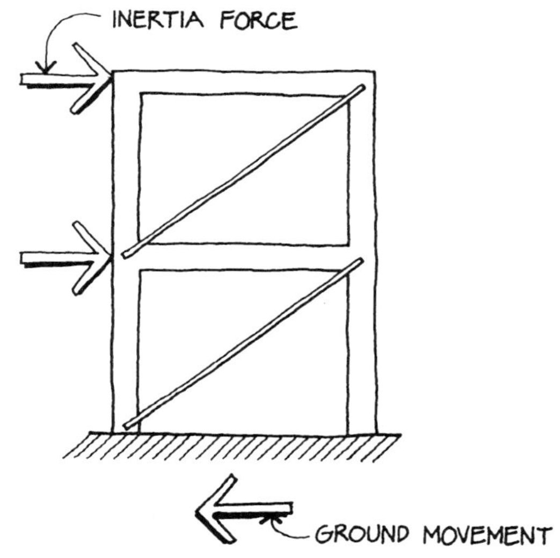

CONCENTRIC DIAGONAL BRACING

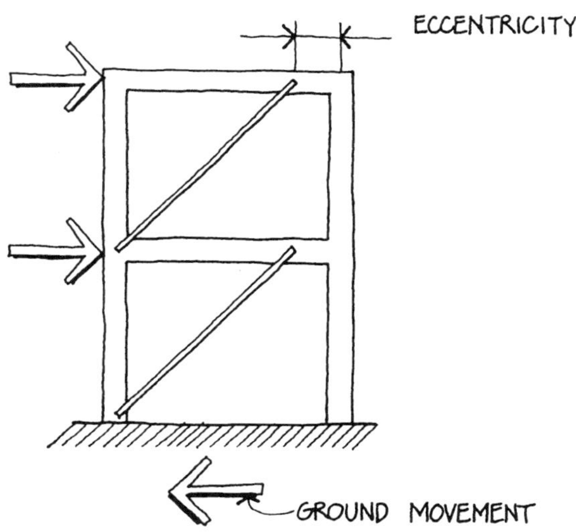

ECCENTRIC DIAGONAL BRACING

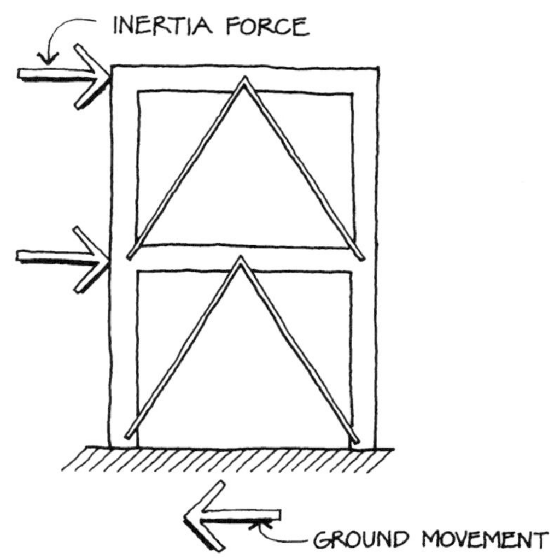

CONCENTRIC CHEVRON BRACING

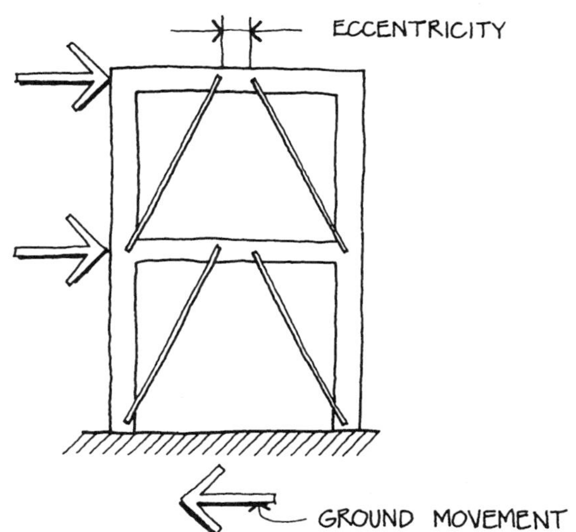

ECCENTRIC CHEVRON BRACING

BRACED FRAMES

TABLE 16-N—STRUCTURAL SYSTEMS

BASIC STRUCTURAL SYSTEM[1]	LATERAL-FORCE-RESISTING SYSTEM—DESCRIPTION	R_W[2]	H[3] × 304.8 for mm
1. Bearing wall system	1. Light-framed walls with shear panels		
	a. Wood structural panel walls for structures three stories or less	8	65
	b. All other light-framed walls	6	65
	2. Shear walls		
	a. Concrete	6	160
	b. Masonry	6	160
	3. Light steel-framed bearing walls with tension-only bracing	4	65
	4. Braced frames where bracing carries gravity loads		
	a. Steel	6	160
	b. Concrete[4]	4	—
	c. Heavy timber	4	65
2. Building frame system	1. Steel eccentrically braced frame (EBF)	10	240
	2. Light-framed walls with shear panels		
	a. Wood structural panel walls for structures three stories or less	9	65
	b. All other light-framed walls	7	65
	3. Shear walls		
	a. Concrete	8	240
	b. Masonry	8	160
	4. Ordinary braced frames		
	a. Steel	8	160
	b. Concrete[4]	8	—
	c. Heavy timber	8	65
	5. Special concentrically braced frames		
	a. Steel	9	240
3. Moment-resisting frame system	1. Special moment-resisting frames (SMRF)		
	a. Steel	12	N.L.
	b. Concrete	12	N.L.
	2. Masonry moment-resisting wall frame	9	160
	3. Concrete intermediate moment-resisting frames (IMRF)[5]	8	—
	4. Ordinary moment-resisting frames (OMRF)		
	a. Steel[6]	6	160
	b. Concrete[7,8]	5	—
4. Dual systems	1. Shear walls		
	a. Concrete with SMRF	12	N.L.
	b. Concrete with steel OMRF	6	160
	c. Concrete with concrete IMRF[5]	9	160
	d. Masonry with SMRF	8	160
	e. Masonry with steel OMRF	6	160
	f. Masonry with concrete IMRF[4]	7	—
	2. Steel EBF		
	a. With steel SMRF	12	N.L.
	b. With steel OMRF	6	160
	3. Ordinary braced frames		
	a. Steel with steel SMRF	10	N.L.
	b. Steel with steel OMRF	6	160
	c. Concrete with concrete SMRF[4]	9	—
	d. Concrete with concrete IMRF[4]	6	—
	4. Special concentrically braced frames		
	a. Steel with steel SMRF	11	N.L.
	b. Steel with steel OMRF	6	160
5. Undefined systems	See Sections 1627.8.3 and 1627.9.2	—	—

N.L.—No limit.
[1]Basic structural systems are defined in Section 1627.6.
[2]See Section 1628.3 for combination of structural system.
[3]H—Height limit applicable to Seismic Zones 3 and 4. See Section 1627.7.
[4]Prohibited in Seismic Zones 3 and 4.
[5]Prohibited in Seismic Zones 3 and 4, except as permitted in Section 1632.2.
[6]Ordinary moment-resisting frames in Seismic Zone 1 meeting the requirements of Section 2211.6 may use an R_w value of 12.
[7]Prohibited in Seismic Zones 2, 3 and 4.
[8]Prohibited in Seismic Zones 2A, 2B, 3 and 4. See Section 1631.2.7.

during an earthquake, their ability to support vertical loads may be eliminated and the structure may collapse. Consequently, bearing wall systems are assigned low values of R_w, varying from 4 to 8. The height limit in Seismic Zones 3 and 4 is either 65 feet or 160 feet.

A *building frame system* is one with an essentially complete frame that provides support for vertical loads. Lateral loads are resisted by shear walls or braced frames, which support no vertical load. The vertical load frame can provide some secondary resistance to lateral loads, and can prevent collapse if the shear walls or braced frames are damaged during an earthquake. Therefore, these systems have intermediate values of R_w, varying from 7 to 10. Height limits in Seismic Zones 3 and 4 vary from 65 to 240 feet.

A *moment-resisting frame system* is one with an essentially complete frame that provides support for vertical loads. Lateral loads are resisted by moment-resisting frames. Because of the great ductility of special moment-resisting frames (SMRF—see page 11), these systems are assigned the maximum R_w value of 12, and their height is unlimited in Seismic Zones 3 and 4.

Ordinary and intermediate moment-resisting frames (OMRF and IMRF) are much less ductile and therefore have lower R_w values, varying from 5 to 8. Of these systems, only OMRF of structural steel are permitted in Seismic Zones 3 and 4, and they are limited to a height of 160 feet.

A *dual system* is one with an essentially complete frame that provides support for vertical loads. Lateral loads are resisted by *both* moment-resisting frames and shear walls or braced frames, in proportion to their relative rigidities. The moment-resisting frame acting independently must be able to resist at least 25 percent of the total required lateral force. These systems have varying degrees of ductility, and their R_W values vary correspondingly, from 6 to 12. For those dual systems that are permitted in Seismic Zones 3 and 4, the height limit is either 160 feet or there is no height limit.

The following examples show how the terms in the basic seismic formula are determined.

Example #1

Denver is located in which seismic zone?

Solution:

This may appear to be so simple that there must be a trick. There is no trick; this kind of question sometimes appears on the actual exam, and is designed to test your understanding of the seismic zone factor Z and your familiarity with the seismic zone map. Simply look at Figure 16-2 in the UBC, which is reproduced on page 8, and you will see that *Denver is in Seismic Zone 1.*

Example #2

Which is more ductile, a building frame system with concrete shear walls, or a moment-resisting frame system with concrete special moment-resisting frames?

Solution:

In general, moment-resisting frames are more ductile than shear walls. Referring to Table 16-N on page 16, the building frame system with concrete shear walls is system 2.3a, which has an R_w value of 8. The moment-resisting frame system with concrete SMRF is system 3.1.b, which has an R_w value of 12. The greater value of R_w indicates greater ductility, and therefore, as expected, *the moment-resisting frame system is more ductile.*

Example #3

An 8-story shear wall building is 100 feet high and 120 feet long. What is its C value?

Solution:

To determine the C value, we first compute the value of the period T.

From page 9, $T = C_t(h_n)^{3/4}$

For a shear wall building, $C_t = 0.020$ ("all other buildings")

h_n = the height of the building = 100 feet

Therefore, $T = 0.020(100)^{3/4} = 0.632$ seconds

From page 7, $C = \frac{1.25S}{T^{2/3}}$

In the absence of more information, soil type S_3 is assumed, which has an S value of 1.5.

Therefore, $C = \frac{1.25(1.5)}{0.632^{2/3}} = 2.55$

Note that raising numbers to the 3/4 or 2/3 power requires the use of a calculator having this function. Also, the number of stories and the length of the building are extraneous and not needed to solve this problem.

Example #4

A three-story apartment building in Seismic Zone 4 has wood-framed bearing walls with plywood shear panels. Its soil conditions and period are not known. What is the base shear as a percentage of the dead load of the building?

Solution:

The base shear $V = \frac{ZIC}{R_w} W$

Z = 0.40 in Seismic Zone 4 (see Table 16-I on page 8).

I = 1.0 since this is not an essential or hazardous facility.

Since we do not know the soil type or building period, we use the maximum value of C = 2.75.

Referring to Table 16-N on page 16, the structure is bearing wall system 1.1.a. Buildings of this type have demonstrated excellent earthquake resistance when properly designed and detailed, and therefore have a relatively high R_w value of 8.

$$V = \frac{ZIC}{R_w} W = \frac{0.40(1.0)(2.75)}{8} W$$
$$= 0.1375 W$$

or *13.75 percent of the dead load.*

Example #5

What is the minimum base shear that may be used in the seismic design of a building in Seismic Zone 4, expressed as a percentage of the building's dead load?

Solution:

We apply the basic seismic formula

$$V = \frac{ZIC}{R_w} W$$

Z = 0.40 in Seismic Zone 4.

To obtain the minimum base shear, we use the minimum value of I = 1.0 and the minimum value of $C/R_w = 0.075$ (see page 9).

$$V = \frac{ZIC}{R_w} W = 0.40\ (1.0)(0.075)\ W = 0.03\ W$$

or *3.0 percent of the dead load.*

DISTRIBUTION OF BASE SHEAR

Once the base shear or total horizontal force has been determined, we must distribute this force to the various levels of the structure. A portion of V, called F_t, is considered concentrated at the top of the structure. $F_t = 0.07\ TV$, except that F_t need not exceed 0.25V and may be considered as zero where T is 0.7 second or less. F_t thus applies to relatively long period buildings to account for a *whiplash* effect that becomes significant in these taller buildings.

The force applied to any level x =

$$F_x = \frac{(V - F_t)w_x h_x}{\Sigma wh}$$

where

w_x = the dead load located at or assigned to level x

h_x = the height in feet above the base to level x

Σwh = the summation of the wh quantities for each level

If the weights of all the stories are equal, the shear will be distributed in the form of a triangle, zero at the base and maximum at the top level, with a horizontal force applied to each floor or roof level above the foundation, as shown above right.

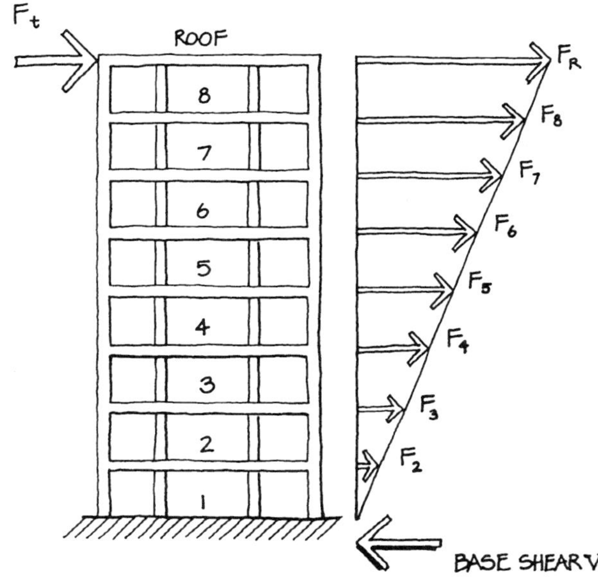

TRIANGULAR DISTRIBUTION OF EARTHQUAKE FORCES

OVERTURNING

Every building must be designed to resist the overturning moments caused by earthquake forces. In computing uplift caused by overturning moment, however, only 85 percent of the dead load may be used to resist uplift forces.

Example #6

A two-story braced frame is subject to the seismic loads and dead loads shown on the following page. What is the overturning moment at the base? For how much uplift should each column be designed?

Solution:

The overturning moment = 20 kips (10 + 10) ft. + 10 kips (10 ft.) = 400 + 100 = 500 *ft.-kips.*

The gross uplift at the left column =

$$\frac{500 \text{ ft.-kips}}{20 \text{ ft.}} = 25 \text{ kips.}$$

This is reduced by 85 percent of the dead load = 0.85(20) = 17.0 kips.

The net uplift is therefore 25 – 17.0 = *8.0 kips.*

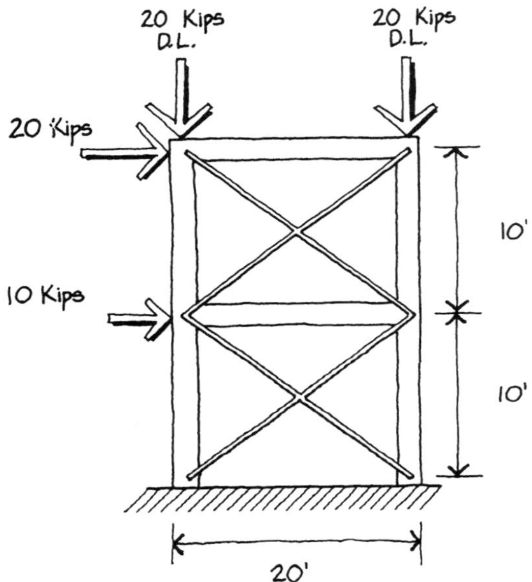

Since seismic loads can come from either direction, each column should be designed to resist this uplift.

Overturning effects are most critical for buildings with a high height-to-width ratio. In other words, a tall, slender building is more susceptible to overturning than a short, squat building.

Top-heavy buildings tend to overturn more than pyramidal buildings.

Heavy equipment, such as water tanks, should be located as low in the structure as possible in order to minimize the overturning effect on the building.

DEFLECTION AND DRIFT

When a building vibrates with the earthquake motion, the structure deflects and the stories move horizontally relative to each other. The story-to-story horizontal movement is called the *story drift*. Stiff systems, such as shear walls and braced frames, have relatively small drifts, while more flexible moment-resisting frames typically have larger drifts.

The UBC limits the amount of this movement in order to insure structural integrity, minimize discomfort to the building's occupants, and restrict damage to brittle nonstructural elements such as glass, plaster walls, etc.

Two buildings next to each other may move differently during an earthquake, and may therefore collide. This collision, called *pounding,* has occurred in many earthquakes and can be disastrous. To minimize the possibility of pounding, a *seismic separation* between the buildings should be provided at least equal to *the sum of the expected drifts* of the two buildings.

Another phenomenon caused by story drift is the *P-Delta effect*. If the story drift due to seismic forces is (delta), and the vertical load in a column is P, then the bending moments in the story are increased by an amount equal to P times Δ or P-Delta. This effect must be taken into account unless it is very low relative to the story bending moments.

DIAPHRAGMS

A diaphragm is the horizontal floor or roof system that distributes lateral forces to the vertical resisting elements, such as shear walls, braced frames, or moment-resisting frames. The action of a diaphragm is similar to that of a horizontal girder spanning between the vertical resisting elements. In this analogy the floor itself (concrete slab, steel deck, plywood sheathing, etc.) may be compared to the girder web that resists shear forces, while the boundary members form the girder flanges or chords, and resist tensile and compressive forces. Although diaphragms are usually considered to be horizontal, they are sometimes inclined, as in a sloping roof.

A *flexible diaphragm* acts as a simple beam in distributing the horizontal forces to the vertical

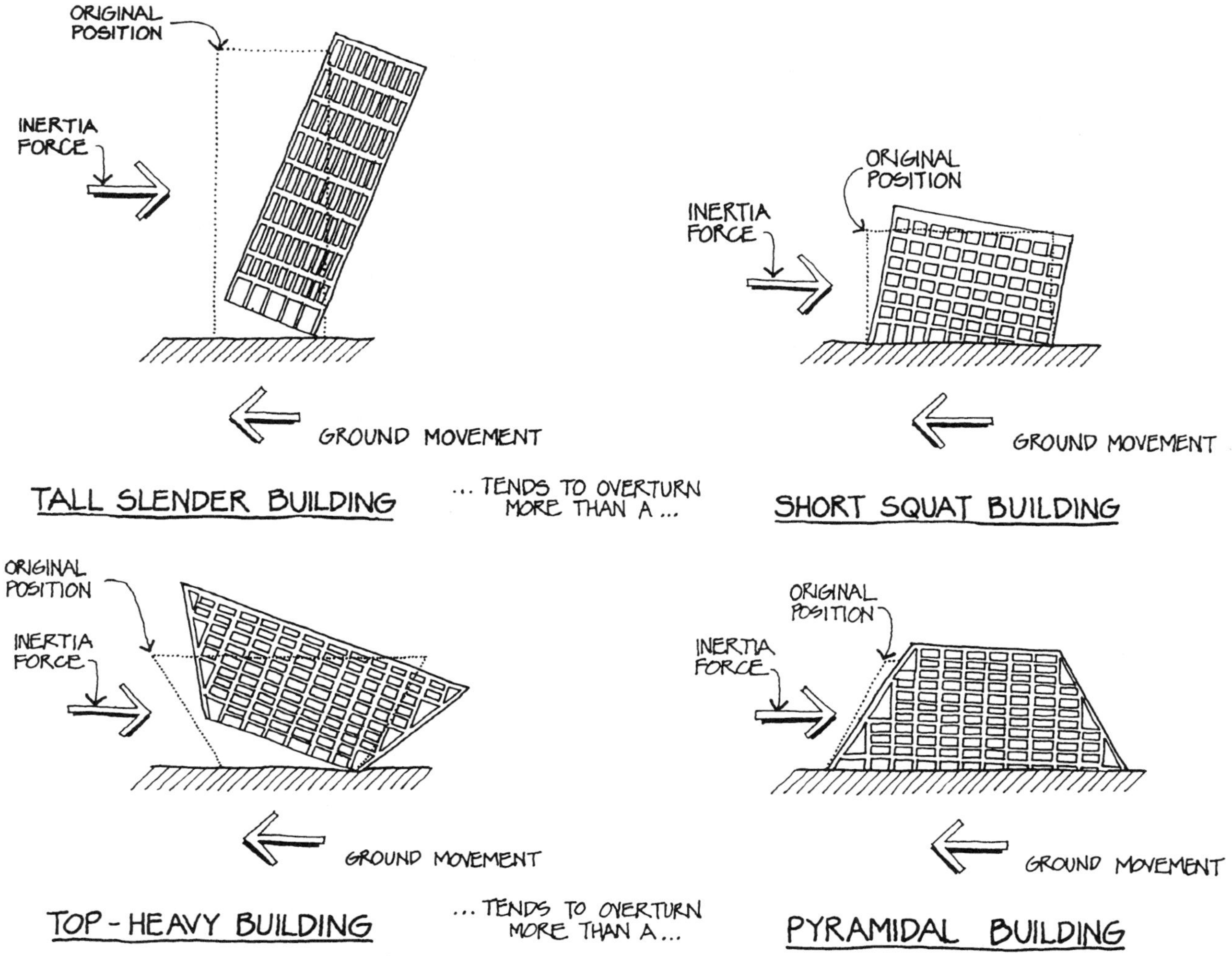

ORIGINAL POSITION
INERTIA FORCE
GROUND MOVEMENT
TALL SLENDER BUILDING
... TENDS TO OVERTURN MORE THAN A ...
ORIGINAL POSITION
INERTIA FORCE
GROUND MOVEMENT
SHORT SQUAT BUILDING
ORIGINAL POSITION
INERTIA FORCE
GROUND MOVEMENT
TOP-HEAVY BUILDING
... TENDS TO OVERTURN MORE THAN A ...
ORIGINAL POSITION
INERTIA FORCE
GROUND MOVEMENT
PYRAMIDAL BUILDING

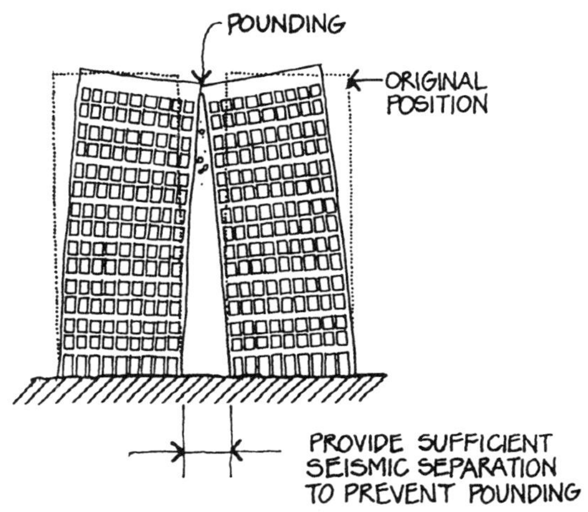

POUNDING
ORIGINAL POSITION
PROVIDE SUFFICIENT SEISMIC SEPARATION TO PREVENT POUNDING
POUNDING EFFECT

resisting elements, which therefore resist the horizontal forces on a tributary area basis. Wood and some steel deck diaphragms are considered flexible.

In the flexible diaphragm shown on page 23, the rigidity values of the walls are not considered. Each end wall resists 1/4 of the lateral load, while the center wall resists 1/2 of the lateral load.

A *rigid diaphragm* distributes the horizontal forces to the vertical resisting elements in proportion to their relative rigidities, rigidity being defined as resistance to deformation. Concrete and some steel deck diaphragms are considered rigid.

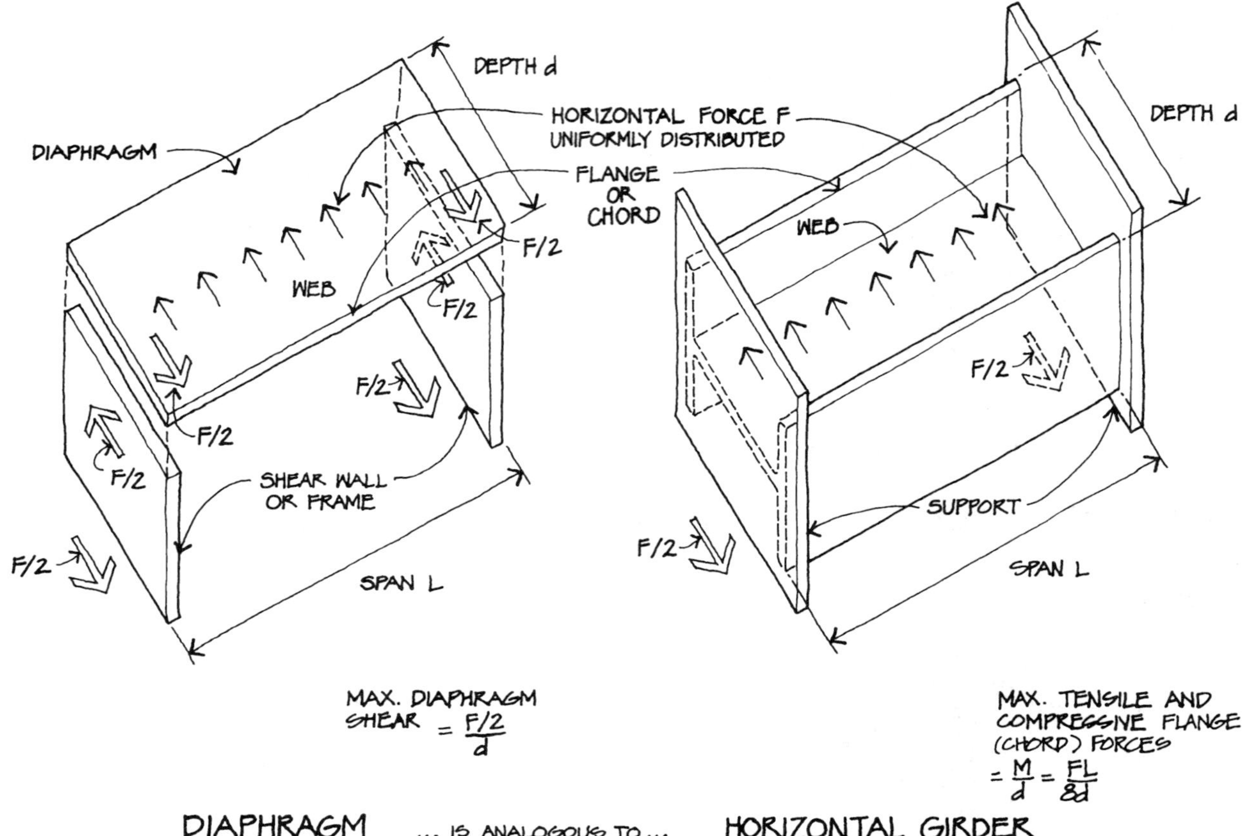

In the rigid diaphragm shown on the following page, the horizontal forces are distributed to the three walls in proportion to their rigidities. $\Sigma R = 2 + 1 + 2 = 5$. Each end wall has 2/5 of the total rigidity and therefore resists 2/5 of the total lateral load. The middle wall has 1/5 of the total rigidity and therefore resists 1/5 of the total lateral load.

Note that the distribution of lateral loads to the shear walls is very different from the distribution with a flexible diaphragm.

The allowable shear values for plywood diaphragms can be found in Table 23-I-J-1 of the UBC, shown on page 24 (IBC Table 2306.4.1, see Appendix). In this table, the term *wood structural panel* refers to various wood-based structural panels, including plywood. Plywood is widely used for diaphragms in small and medium-sized buildings. It is lightweight, economical, quickly and easily installed, and moderately strong. As shown in the table, its shear value depends on the plywood thickness, the size and spacing of the nails, and whether blocking is used at the edges of the plywood panels (a higher value is allowed where blocking is used). Adequate nailing must always be provided around the perimeter of the plywood diaphragm.

For example, the allowable shear in a 3/8" plywood blocked diaphragm of Structural I grade with 8d nails spaced at 6" at plywood panel edges and 3" framing members is 300 pounds per foot for seismic loading.

Diaphragms of steel deck have shear values varying from about 100 pounds per lineal foot to 2,600

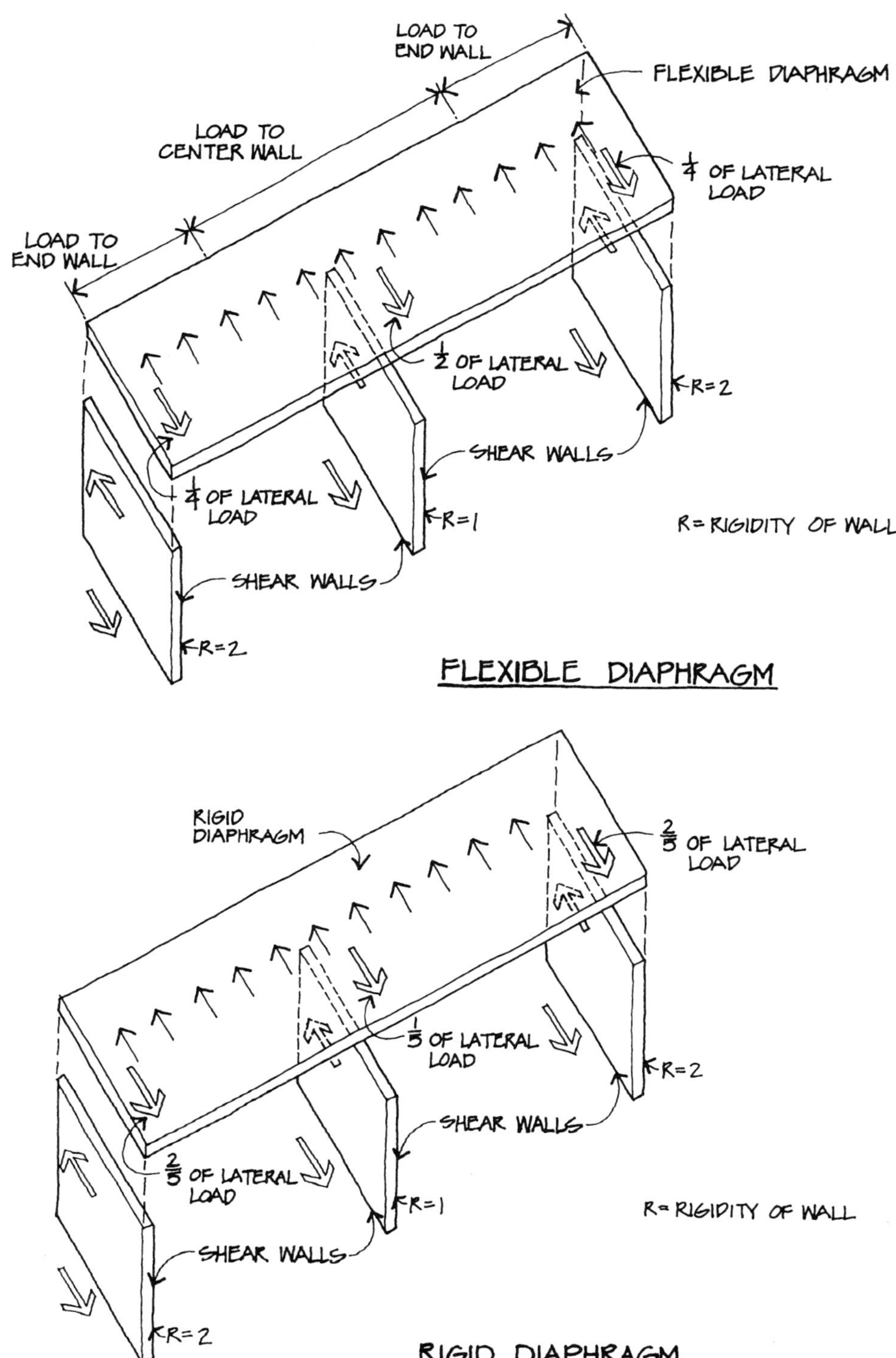
LOAD TO END WALL
FLEXIBLE DIAPHRAGM
LOAD TO CENTER WALL
¼ OF LATERAL LOAD
LOAD TO END WALL
½ OF LATERAL LOAD
R=2
SHEAR WALLS
¼ OF LATERAL LOAD
R=1
R=RIGIDITY OF WALL
SHEAR WALLS
R=2
FLEXIBLE DIAPHRAGM
RIGID DIAPHRAGM
⅖ OF LATERAL LOAD
⅕ OF LATERAL LOAD
R=2
SHEAR WALLS
⅖ OF LATERAL LOAD
R=1
R=RIGIDITY OF WALL
SHEAR WALLS
R=2
RIGID DIAPHRAGM

TABLE 23-I-J-1 — ALLOWABLE SHEAR IN POUNDS PER FOOT FOR HORIZONTAL WOOD STRUCTURAL PANEL DIAPHRAGMS WITH FRAMING OF DOUGLAS FIR-LARCH OR SOUTHERN PINE

PANEL GRADE	COMMON NAIL SIZE	MINIMUM NAIL PENETRATION IN FRAMING (inches)	MINIMUM NOMINAL PANEL THICKNESS (inches)	MINIMUM NOMINAL WIDTH OF FRAMING MEMBER (inches)	BLOCKED DIAPHRAGMS: Nail spacing (in.) at diaphragm boundaries (all cases), at continuous panel edges parallel to load (Cases 3 and 4) and at all panel edges (Cases 5 and 6) × 25.4 for mm				UNBLOCKED DIAPHRAGMS: Nails spaced 6" (152 mm) max. at supported edges	
					6	4	$2^1/_2$ [2]	2 [2]	Case 1 (No unblocked edges or continuous joints parallel to load)	All other configurations (Cases 2, 3, 4, 5 and 6)
					Nail spacing (in.) at other panel edges × 25.4 for mm					
		× 25.4 for mm			6	6	4	3		
					× 0.0146 for N/mm					
Structural I	6d	$1^1/_4$	$^5/_{16}$	2	185	250	375	420	165	125
				3	210	280	420	475	185	140
	8d	$1^1/_2$	$^3/_8$	2	270	360	530	600	240	180
				3	300	400	600	675	265	200
	10d[3]	$1^5/_8$	$^{15}/_{32}$	2	320	425	640	730	285	215
				3	360	480	720	820	320	240
C-D, C-C, Sheathing, and other grades covered in U.B.C. Standard 23-3 or 23-9	6d	$1^1/_4$	$^5/_{16}$	2	170	225	335	380	150	110
				3	190	250	380	430	170	125
			$^3/_8$	2	185	250	375	420	165	125
				3	210	280	420	475	185	140
	8d	$1^1/_2$	$^3/_8$	2	240	320	480	545	215	160
				3	270	360	540	610	240	180
			$^7/_{16}$	2	255	340	505	575	230	170
				3	285	380	570	645	255	190
			$^{15}/_{32}$	2	270	360	530	600	240	180
				3	300	400	600	675	265	200
	10d[3]	$1^5/_8$	$^{15}/_{32}$	2	290	385	575	655	255	190
				3	325	430	650	735	290	215
			$^{19}/_{32}$	2	320	425	640	730	285	215
				3	360	480	720	820	320	240

pounds per lineal foot, depending on the steel deck thickness, the size and spacing of welds or other attachments between the deck and the framing and between deck sheets, and the deck span. The shear value of steel deck can also be increased by placing concrete fill over the steel deck.

Concrete slabs can also be used as diaphragms, and a six-inch concrete slab, for example, has an allowable shear of about 10,000 pounds per lineal foot.

Although most structural decking materials, such as plywood, steel deck, and concrete, can function as diaphragms, some cannot. Among these are straight tongue-and-groove sheathing and light gauge corrugated metal.

In such cases, alternate methods, such as a horizontal bracing system, must be used to distribute the lateral forces to the vertical resisting elements.

Another factor that must be considered in diaphragm design is the horizontal deflection of the diaphragm. Referring again to the girder analogy, as the span of the diaphragm between vertical resisting elements increases, the deflection of the diaphragm increases at a greater rate. If the diaphragm deflection becomes too great, there can be excessive damage, and possibly even collapse, of elements laterally supported by the diaphragm. Accordingly, the UBC limits the span-to-depth ratio of plywood diaphragms to 4:1, and further requires that the diaphragm deflection not exceed that for which the diaphragm and any attached resisting elements will maintain their structural integrity without danger to the occupants of the structure.

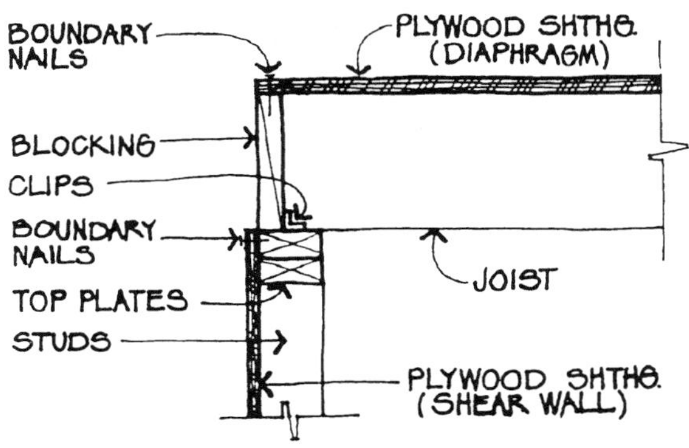

WOOD JOISTS PERPENDICULAR TO WOOD STUD WALL

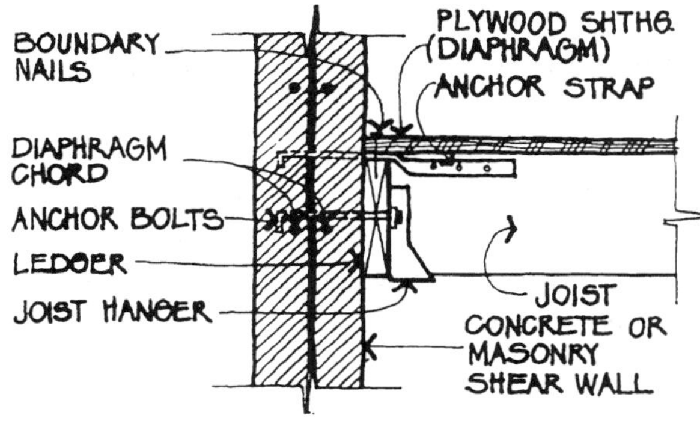

WOOD JOISTS PERPENDICULAR TO CONCRETE OR MASONRY WALL

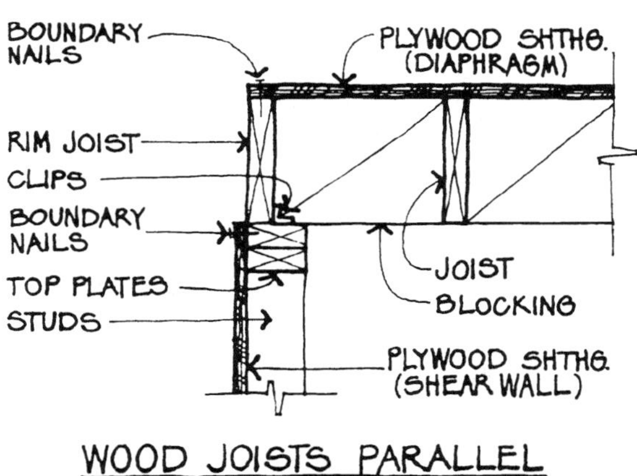

WOOD JOISTS PARALLEL TO WOOD STUD WALL

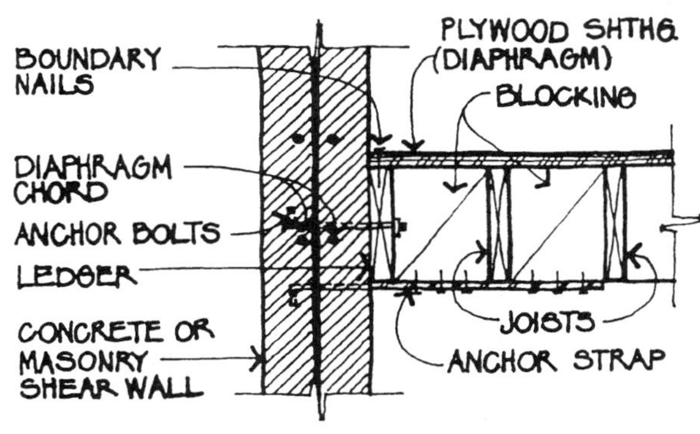

WOOD JOISTS PARALLEL TO CONCRETE OR MASONRY WALL

Diaphragms must be adequately connected to the vertical resisting elements, such as shear walls. The connections must be designed to do the following:

1. Transfer the diaphragm shear force, which is parallel to the shear wall, into the wall.
2. Connect the diaphragm to the chord, or flange.
3. Connect the wall to the diaphragm for seismic forces perpendicular to the wall.

Examples of several diaphragm connection details are shown above. Although there is an infinite variety of diaphragm connection details, if you understand the concepts of those shown, it will help you answer any questions on this subject which may appear on the exam.

In the two details above, the diaphragm shear is transferred to the shear wall as follows: from the diaphragm (plywood sheathing) through the roof boundary nails to the rim joist or blocking, then through the clips to the top plates, then through the wall boundary nails to the shear wall (plywood wall sheathing).

The chord is either the rim joist or the top plates, connected to the diaphragm as described above.

Seismic or wind forces perpendicular to the wall are small and can be transferred from the top plates to the joists or blocking through toenails, and then through nails to the diaphragm.

In the two details in the right column on page 25, the diaphragm shear is transferred from the sheathing through the boundary nails to the ledger, and then from the ledger to the wall through the anchor bolts.

The diaphragm chord consists of a couple of reinforcing bars in the wall, and the connection of the diaphragm to the wall is as described above.

Seismic forces perpendicular to the wall are transferred through the anchor straps to the joists or blocking, and then through nails to the sheathing.

The joist hangers support the joists for vertical load.

In the detail below, the diaphragm shear is transferred from the steel deck through the welds to the continuous angle, then through the anchor bolts to the wall.

A couple of reinforcing bars in the wall form the diaphragm chord, and the connection of the diaphragm to the wall is as described above.

Seismic forces perpendicular to the wall are transferred through the dowels into the concrete fill.

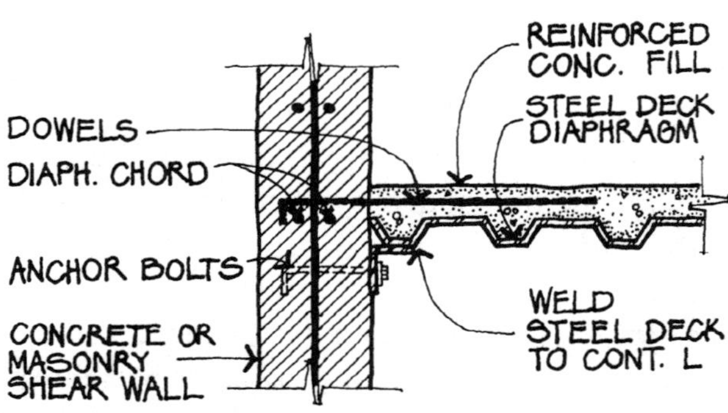

STEEL DECK CONNECTION TO CONCRETE OR MASONRY WALL

In the detail below, the diaphragm shear is transferred from the steel deck through the welds to the continuous plate, then through the welded studs to the wall. Seismic forces perpendicular to the wall are transferred to the steel deck diaphragm in the same way. Since this type of detail would probably be used over an interior wall, not at the diaphragm boundary, a diaphragm chord is not required.

COLLECTORS

Where a shear resisting element is discontinuous, a collector member, often called a *strut* or *drag strut*, is used to collect seismic load from the diaphragm to which it is attached and deliver it to the shear resisting element. Two examples are shown on the following page.

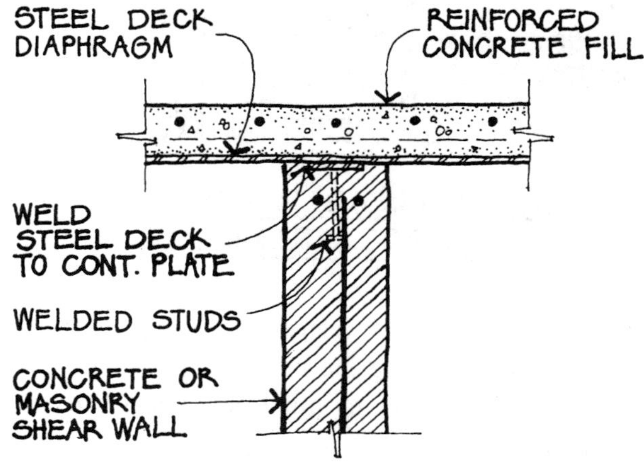

STEEL DECK CONNECTION TO CONCRETE OR MASONRY WALL BELOW

TORSION

Torsion is the rotation caused in a diaphragm when the center of mass (where the resultant load is applied) does not coincide with the center of rigidity (where the resultant load is resisted),

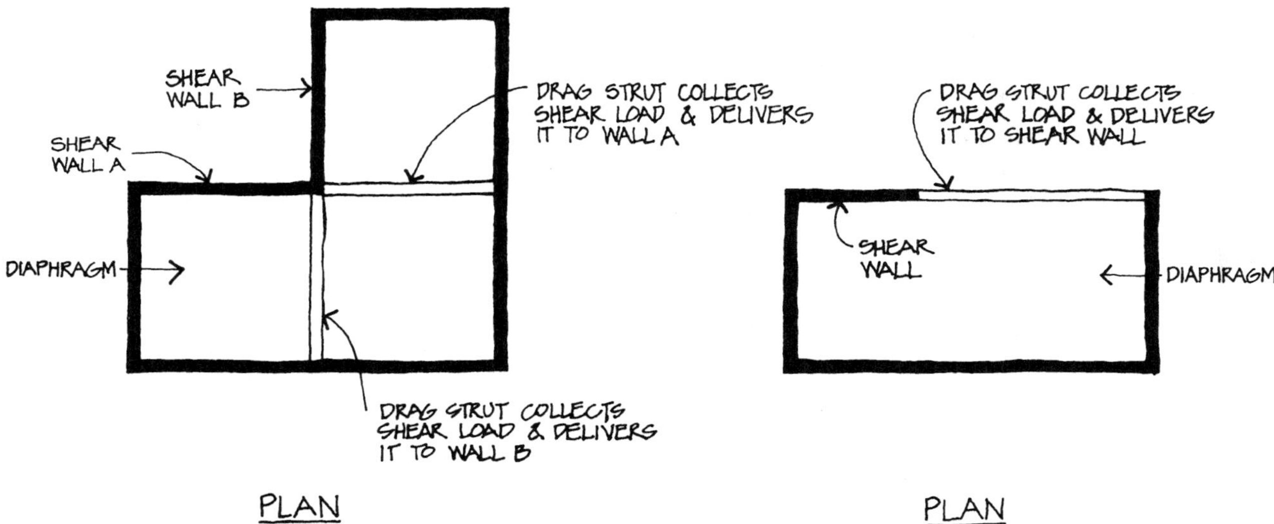

COLLECTORS

as shown in the right column of this page. Torsion occurs only in rigid diaphragms.

The torsional moment causes forces in the shear resisting elements in direct proportion to (1) the magnitude of the torsional moment, (2) the distance of the shear resisting element from the center of rigidity, and (3) the rigidity of the shear resisting element.

The UBC requires that when the force caused by torsional moment is in the same direction as the force caused by the applied seismic force, the forces must be added. However, when the force caused by torsional moment is in the opposite direction from that caused by the applied seismic force, it is not subtracted.

In addition, the code requires that an arbitrary amount of *accidental torsion* must be provided for in the design, even if the building is completely symmetrical. This is to account for conditions such as nonuniform distribution of vertical loads, unsymmetrical location of floor openings, eccentricity in rigidity caused by non-structural elements, and seismic ground motion that may include a torsional component.

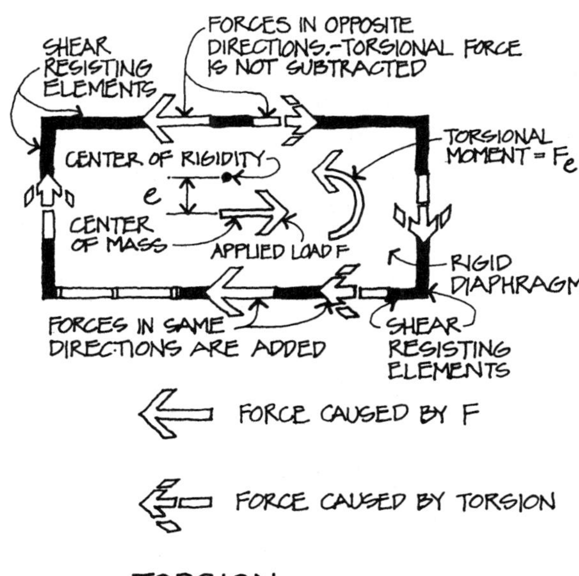

TORSION

To provide for this accidental torsion, the mass at each level is assumed to be displaced a distance equal to 5 percent of the building dimension at that level perpendicular to the direction of the force, as shown on page 28.

Thus, in the north-south direction, the accidental torsional moment is equal to V times .05a, and in the east-west direction, it is equal to V times .05b.

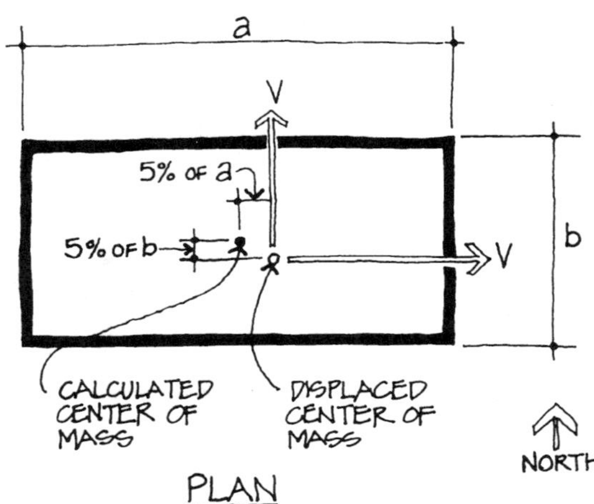

ACCIDENTAL TORSION

Torsional effects are most significant in unsymmetrical buildings. These include buildings with an L configuration and structures whose shear resisting elements are in a core far from the building's perimeter, particularly where the core is not near the center of the building. While torsional effects cannot be entirely eliminated, they can be reduced by locating shear resisting elements at the perimeter of a structure, and by making the building and its system of shear resisting elements symmetrical. A more complete discussion of symmetry and regularity is found starting on page 29.

PARTS OF STRUCTURES

The UBC requires that parts of structures and their attachments must be designed for lateral forces in accordance with the formula

$F_p = ZIC_pW_p$

where

- F_p = the lateral seismic force on the part of the structure under consideration
- Z and I = the seismic zone factor and the importance factor used for the building
- C_p = a numerical coefficient whose value is given in Table 16-O, reproduced on the following page
- W_p = the weight of the part of the structure under consideration

Example #7

An 8"-thick concrete parapet wall in Seismic Zone 4 extends four feet above the roof line. For what moment should the wall be designed? Assume I is equal to 1.0.

Solution:

$F_p = ZIC_pW_p$

Z = 0.40 in Seismic Zone 4, I is equal to 1.0 according to the problem statement, $C_p = 2.0$ for cantilevered parapet walls according to Table 16-O, and W_p = 8/12 x 150 = 100 pounds per square foot.

$F_p = 0.40 \times 1 \times 2.0 \times 100 = 80$ pounds per square foot.

Since F_p is equal to 80 pounds per square foot and the wall weighs 100 pounds per square foot, we can say that the parapet wall has an 80 percent seismic factor; in other words, it is designed for a horizontal force perpendicular to its surface equal to 80 percent of its own weight.

The moment = 80 × 4 × 4/2 = *640'#*

COMBINED VERTICAL AND HORIZONTAL FORCES

The UBC requires that all building components be designed for the following load combinations:

1. Dead plus floor live plus seismic.
2. Dead plus floor live plus snow + seismic.

TABLE 16-O—HORIZONTAL FORCE FACTOR, CP

	ELEMENTS OF STRUCTURES, NONSTRUCTURAL COMPONENTS AND EQUIPMENT[1]	VALUE OF C_p	FOOTNOTE
1.	**Elements of structures**		
	1. Walls including the following:		
	a. Unbraced (cantilevered) parapets	2.00	
	b. Other exterior walls above the ground floor	0.75	2,3
	c. All interior bearing and nonbearing walls and partitions	0.75	3
	d. Masonry or concrete fences over 6 feet (1829 mm) high	0.75	
	2. Penthouse (except when framed by an extension of the structural frame)	0.75	
	3. Connections for prefabricated structural elements other than walls, with force applied at center of gravity	0.75	4
	4. Diaphragms	—	5
2.	**Nonstructural components**		
	1. Exterior and interior ornamentations and appendages	2.00	
	2. Chimneys, stacks, trussed towers and tanks on legs:		
	a. Supported on or projecting as an unbraced cantilever above the roof more than one half their total height	2.00	
	b. All others, including those supported below the roof with unbraced		
	projection above the roof less than one half its	0.75	
	height, or braced or guyed to the structural frame at or above their centers of	2.00	
	mass	0.75	10
	3. Signs and billboards		
	4. Storage racks (include contents)	0.75	
	5. Anchorage for permanent floor-supported cabinets and book stacks more than	0.75	4,6,7,11
	5 feet (1524 mm) in height (include contents)	0.75	4,9
	6. Anchorage for suspended ceilings and light fixtures		
	7. Access floor systems		
3.	**Equipment**		
	1. Tanks and vessels (include contents), including support systems and anchorage	0.75	
	2. Electrical, mechanical and plumbing equipment and associated conduit, ductwork and piping, and machinery	0.75	8

Note that the roof live load and wind load are not considered to act concurrently with the seismic load, since the likelihood that they will be acting on the structure during an earthquake is very remote.

Allowable stresses may be increased one-third when considering earthquake forces either acting alone or combined with the vertical loads above.

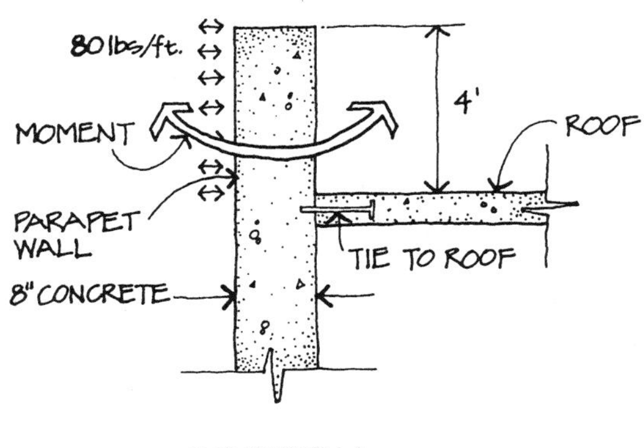

REGULAR AND IRREGULAR STRUCTURES

According to the UBC, structures are designated as structurally regular or irregular. A regular structure has no significant discontinuities in plan, vertical configuration, or lateral force resisting systems. An irregular structure, on the other hand, has significant discontinuities such as those in Table 16-L (vertical irregularities) or Table 16-M (plan irregularities), reproduced on page 30 (IBC Tables 1616.5.1.1 and 1616.5.1.2, see Appendix).

Regular and symmetrical structures exhibit more favorable and predictable seismic response characteristics than irregular structures. Therefore, the use of irregular structures in earthquake-prone areas should be avoided if possible.

TABLE 16-L—VERTICAL STRUCTURAL IRREGULARITIES

IRREGULARITY TYPE AND DEFINITION	REFERENCE SECTION
1. **Stiffness irregularity—soft story** A soft story is one in which the lateral stiffness is less than 70 percent of that in the story above or less than 80 percent of the average stiffness of the three stories above.	1627.8.3, Item 2
2. **Weight (mass) irregularity** Mass irregularity shall be considered to exist where the effective mass of any story is more than 150 percent of the effective mass of an adjacent story. A roof which is lighter than the floor below need not be considered.	1627.8.3, Item 2
3. **Vertical geometric irregularity** Vertical geometric irregularity shall be considered to exist where the horizontal dimension of the lateral force-resisting system in any story is more than 130 percent of that in an adjacent story. One-story penthouses need not be considered.	1627.8.3, Item 2
4. **In-plane discontinuity in vertical lateral-force-resisting element** An in-plane offset of the lateral load-resisting elements greater than the length of those elements.	1628.7
5. **Discontinuity in capacity—weak story** A weak story is one in which the story strength is less than 80 percent of that in the story above. The story strength is the total strength of all seismic-resisting elements sharing the story shear for the direction under consideration.	1627.9.1

TABLE 16-M—PLAN STRUCTURAL IRREGULARITIES

IRREGULARITY TYPE AND DEFINITION	REFERENCE SECTION
1. **Torsional irregularity—to be considered when diaphragms are not flexible** Torsional irregularity shall be considered to exist when the maximum story drift, computed including accidental torsion, at one end of the structure transverse to an axis is more than 1.2 times the average of the story drifts of the two ends of the structure.	1631.2.9, Item 6
2. **Reentrant corners** Plan configurations of a structure and its lateral force-resisting system contain reentrant corners, where both projections of the structure beyond a reentrant corner are greater than 15 percent of the plan dimension of the structure in the given direction.	1631.2.9, Items 6 and 7
3. **Diaphragm discontinuity** Diaphragms with abrupt discontinuities or variations in stiffness, including those having cutout or open areas greater than 50 percent of the gross enclosed area of the diaphragm, or changes in effective diaphragm stiffness of more than 50 percent from one story to the next.	1631.2.9, Item 6
4. **Out-of-plane offsets** Discontinuities in a lateral force path, such as out-of-plane offsets of the vertical elements.	1628.7; 1631.2.9, Item 6; 2211.8
5. **Nonparallel systems** The vertical lateral load-resisting elements are not parallel to or symmetric about the major orthogonal axes of the lateral force-resisting system.	1631.1

However, the UBC does not prohibit irregular structures. Instead, it contains specific design requirements for each type of irregularity. In some cases of irregularity, the static lateral force procedure as described in this lesson is not permitted, and a dynamic procedure is required.

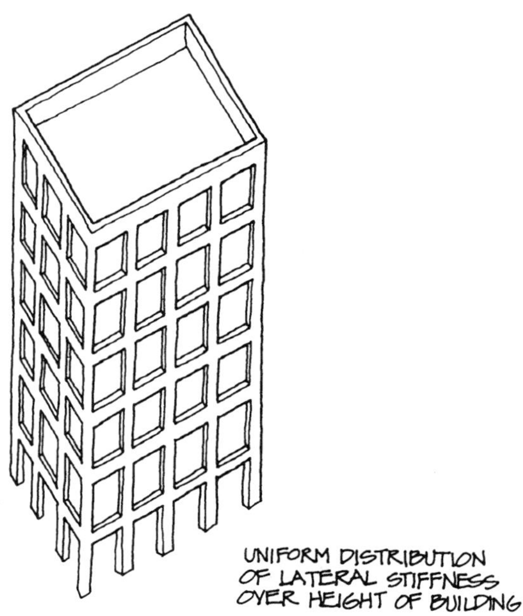

STIFFNESS REGULARITY

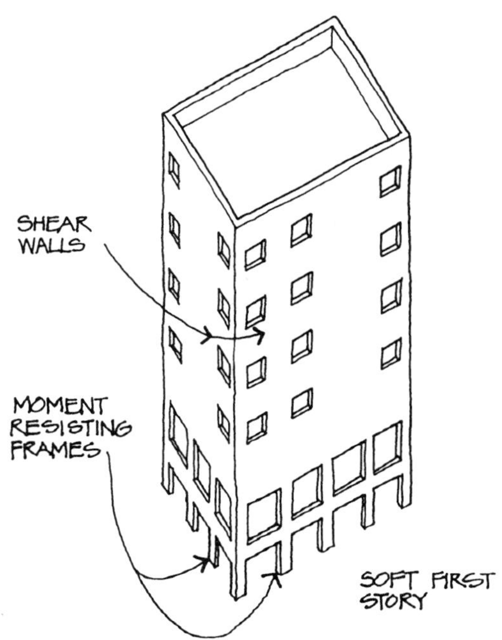

STIFFNESS IRREGULARITY

The following is a summary of some recommended design practices concerning structural regularity.

1. Structures should be regular in stiffness and geometry, both in plan and elevation.
2. Abrupt changes of shape, stiffness, or resistance, such as a *soft* first story, should be avoided.
3. Portions of a building that are different in size, shape, or rigidity should have a seismic separation, as shown below right.
4. Reentrant corners should be avoided. Such corners occur in L-, T-, U-, and cross-shaped plans.
5. If a reentrant corner is unavoidable, it should be strengthened by using drag struts (see page 27), or, preferably, a seismic separation should be provided, as shown on the following page.
6. Torsion should be minimized by making the building symmetrical and regular in geometry and stiffness, and by providing lateral load resisting elements at the building's perimeter as shown on page 33.

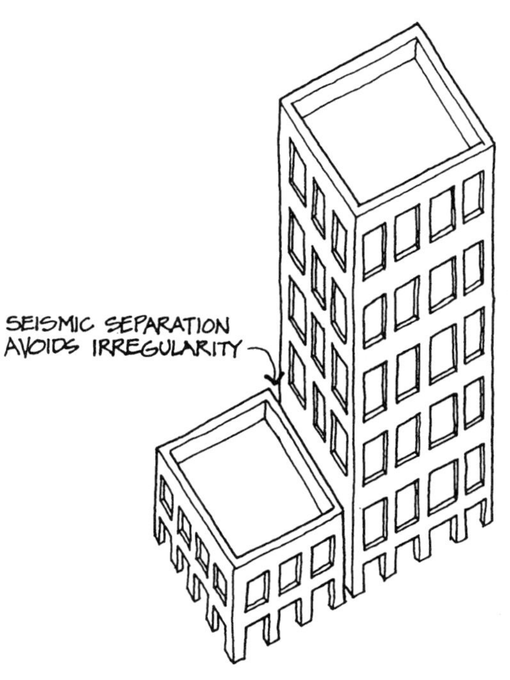

VERTICAL GEOMETRIC IRREGULARITY

Distortion of the soft story, Olive View Hospital, San Fernando earthquake, 1971 (N.O.A.A. photograph)

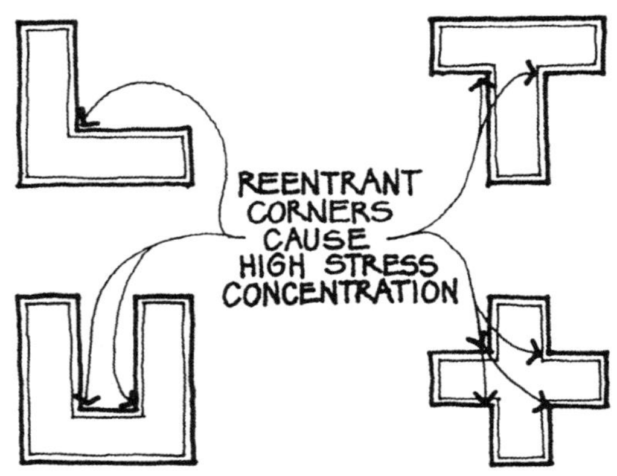

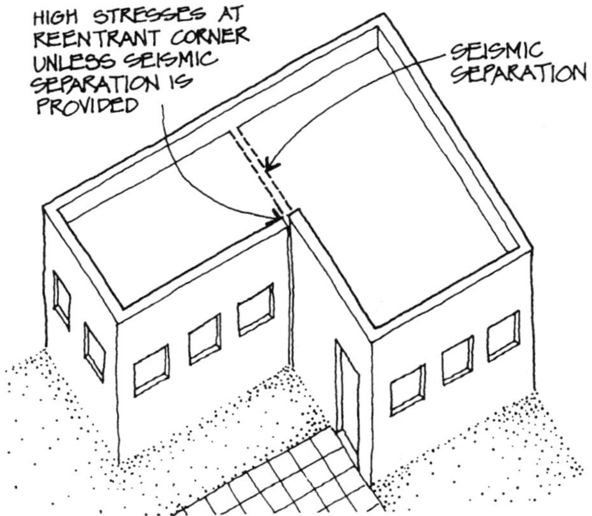

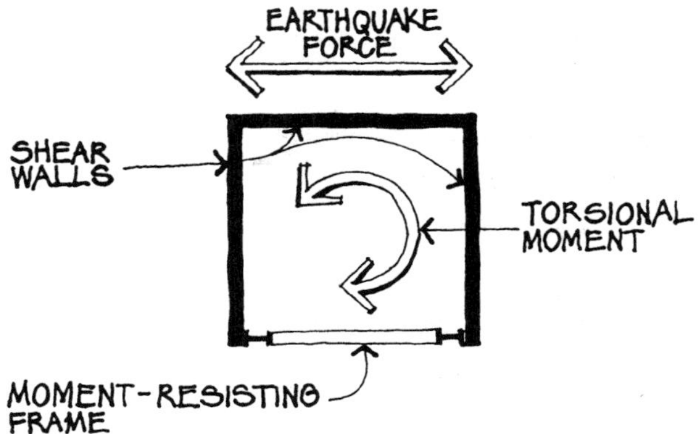

7. Open-front buildings have high torsional stresses because the rear wall, which generally has minimal openings, is very rigid, while the front is very flexible. The most practical solution is to make the front as rigid as possible, by providing a moment-resisting frame, as shown on the right.
8. Complex or asymmetrical building shapes, which may introduce stress concentrations and/or torsion, should be avoided. If necessary, seismic separations should be provided.
9. There should be a direct path for force transfer. Shear walls or other lateral load resisting elements should be continuous to the foundation. In-plane or out-of-plane offsets should be avoided.
10. Diaphragms with abrupt discontinuities, as shown on the top right of page 33, should be avoided.
11. Column stiffness variation, as shown on page 34, should be avoided, if possible.

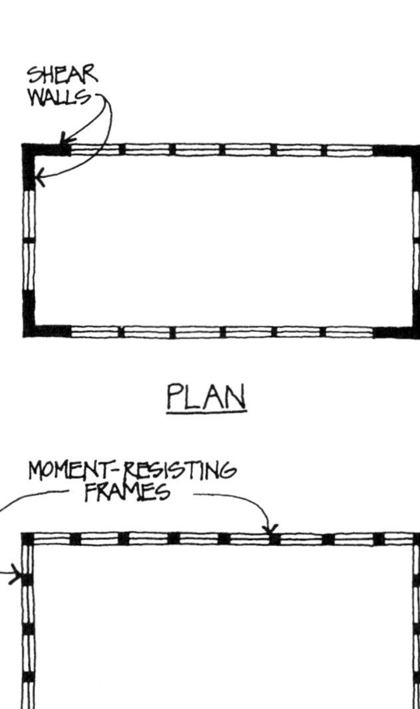

PLAN

TORSIONAL REGULARITY

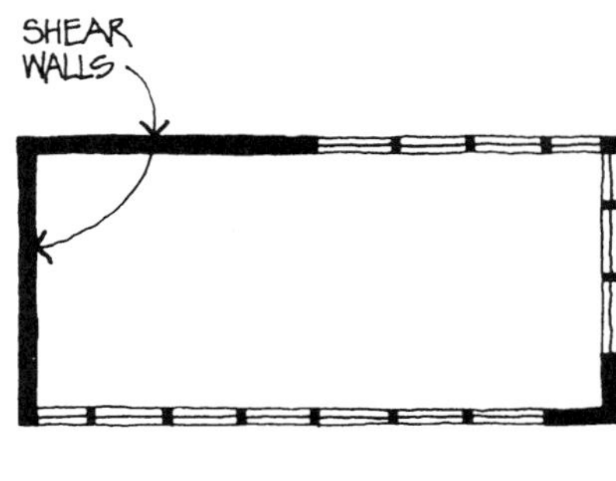

PLAN

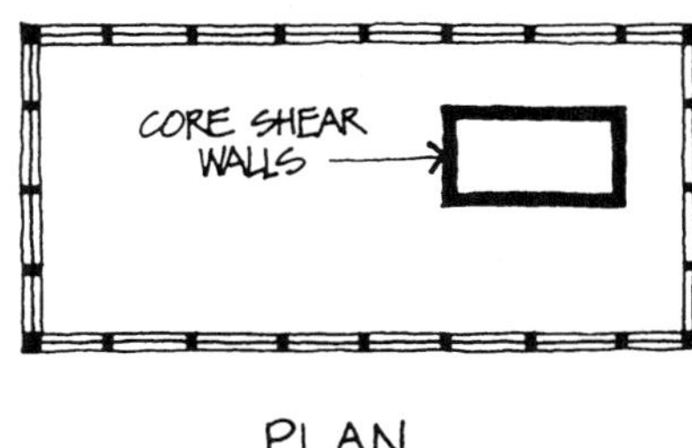

PLAN

TORSIONAL IRREGULARITY

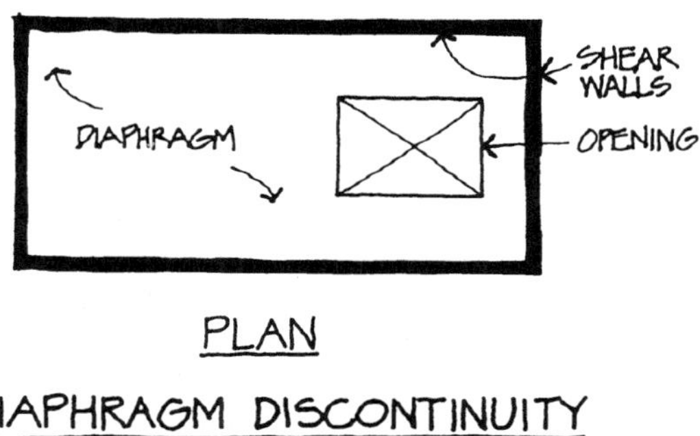

PLAN

DIAPHRAGM DISCONTINUITY

BASE ISOLATION

In all of the preceding discussion, the structure is in direct contact with the ground. Therefore, during an earthquake, the ground motion is transmitted directly to the structure.

A new approach, which has been used on a few buildings, is base isolation. In this method, the structure is isolated from the ground by specially-designed bearings and dampers that absorb earthquake forces, thus reducing the building's acceleration from earthquake ground motion. The acceptance of base isolation as an alternative design approach is likely to increase as we gain more experience in its use.

TUBULAR CONCEPT

As previously discussed, special moment-resisting frames (see page 17) have generally performed well in earthquakes because of their ductility. However, for very tall buildings, this system may not be economical.

A relatively recent development, pioneered by the late Fazlur Khan of S.O.M., is the concept of tubular behavior for tall buildings subject to lateral loads from earthquake or wind. This consists of closely-spaced columns at the perimeter of the building connected by deep spandrel beams at each floor to form, in effect, a

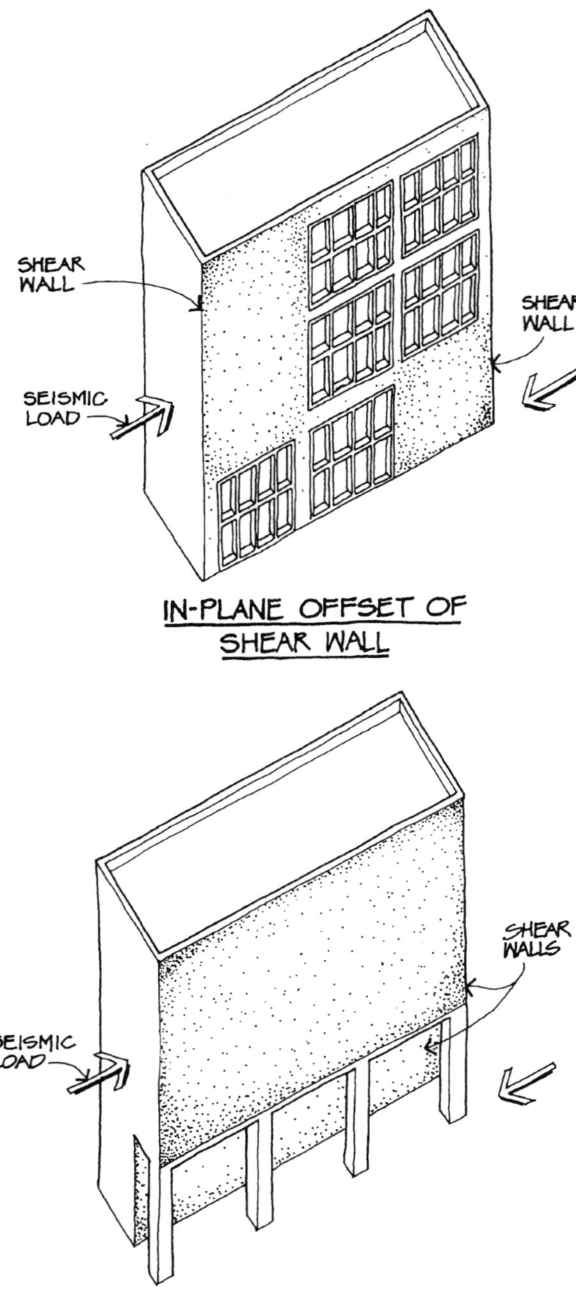

IN-PLANE OFFSET OF SHEAR WALL

OUT-OF-PLANE OFFSET OF SHEAR WALL

perforated wall at each facade. This system behaves like a tube, or box beam, that cantilevers from the ground when subject to lateral earthquake or wind forces.

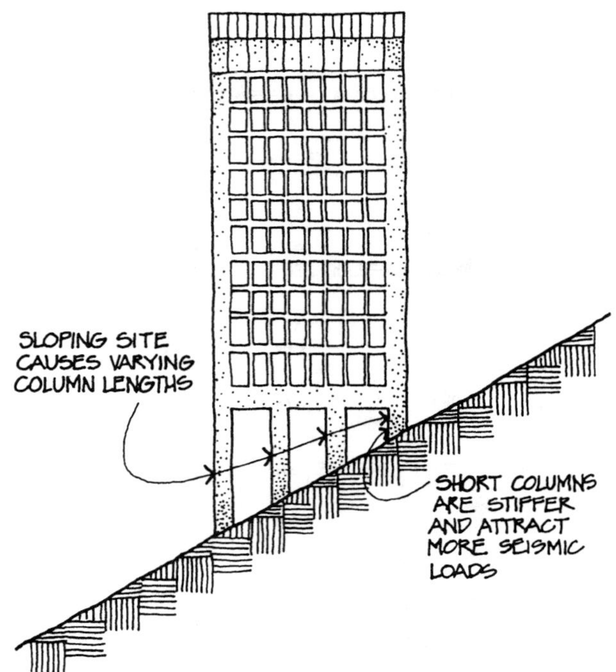

COLUMN STIFFNESS IRREGULARITY

Tubular systems can provide strength and stiffness economically, particularly for tall buildings. Some of the tallest buildings constructed in the United States employ tubular systems, including the John Hancock Building and Sears Tower, both in Chicago, and the former twin towers of the World Trade Center in New York.

NONSTRUCTURAL CONSIDERATIONS

The primary goals of earthquake design are to prevent building collapse and loss of life. Even if a building doesn't collapse, however, its occupants are subject to injury from falling objects, such as equipment cabinets, bookshelves, suspended ceilings, lighting fixtures, partitions, etc. Doors may become jammed in their frames, thus becoming difficult or impossible to force open. Falling glass, parapets, and other elements pose a hazard to people both inside and outside the building. Nonstructural building elements must

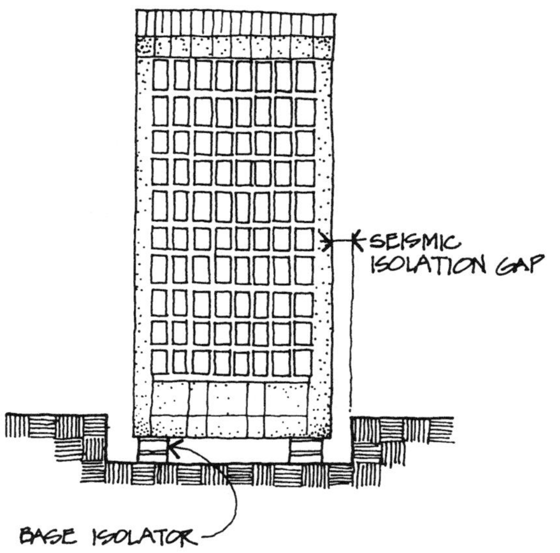

BASE ISOLATION

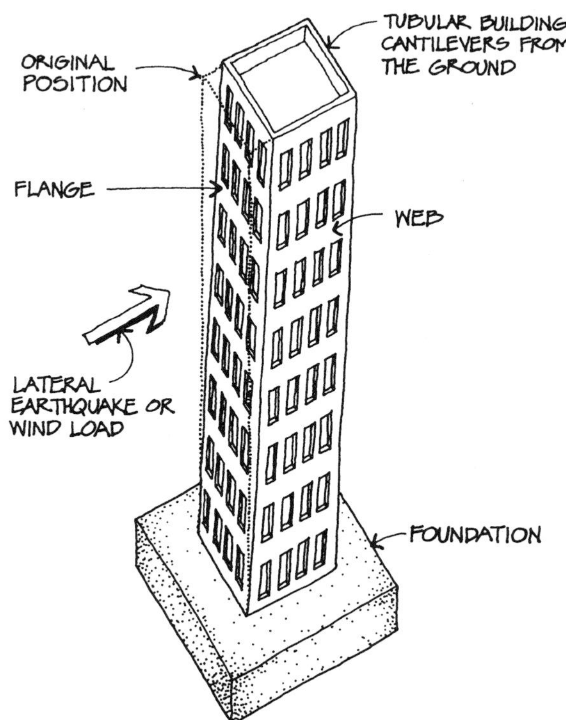

TUBULAR CONCEPT

therefore be designed with these dangers in mind.

The building and its occupants may be subject to secondary hazards, such as fire and flooding. Alarm and sprinkler systems are vulnerable to damage and should therefore be designed to be operable following an earthquake.

Evacuation of a building following an earthquake may be difficult or even impossible because of debris, lack of light, or flooding.

Elevators are particularly vulnerable to damage. Therefore, the design should carefully consider all exit paths.

Mechanical and electrical systems should be designed to survive an earthquake, if possible, so that the building may be quickly returned to service. This is especially true for hospitals and other emergency centers.

It is not enough to design a building so that it can survive a major earthquake. All of its components, elements, and systems must be designed and braced so that damage during and after an earthquake is minimized. Sketches of a couple of typical details for nonstructural elements are shown on the following page.

EXISTING BUILDINGS

Sometimes, an architect must evaluate the seismic resistance of an existing building. Whether a building is new or existing, the same seismic concepts apply. Thus, regularity, continuity, redundancy, and other seismic design practices discussed in this lesson are pertinent to existing buildings.

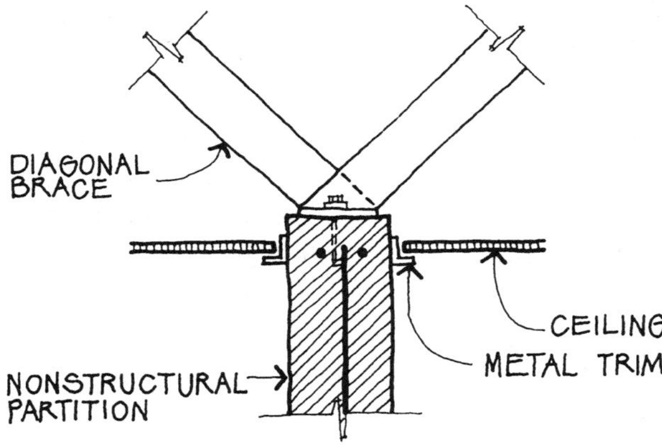

TYPICAL NONSTRUCTURAL PARTITION DETAIL

Some general considerations in the seismic evaluation of existing buildings are as follows:

1. There should be a complete lateral force resisting system, and all of its elements including connections, anchorages, and diaphragms should be adequate to resist the seismic loads.
2. Significant corrosion, deterioration, or decay of the structural materials used in the vertical and lateral force resisting systems can impair their capacity and may require corrective action.
3. Chimneys, cornices, and parapets should be adequately reinforced and anchored to prevent them from breaking off and falling.
4. Exterior cladding, veneer, and wall panels should be adequately attached to the framing to prevent them from falling off.
5. Large openings in diaphragms should be reinforced so that the diaphragm has sufficient strength and stiffness to resist seismic loads.
6. Walls with garage doors or other large openings may be inadequate to resist seismic loads and may undergo distress or collapse unless stiffened or reinforced.
7. Glazing should be sufficiently isolated from the structure to prevent it from shattering and falling out.
8. Wood frame buildings generally perform well in an earthquake and do not pose a significant threat to life. However, they may be damaged and their contents may be badly shaken.

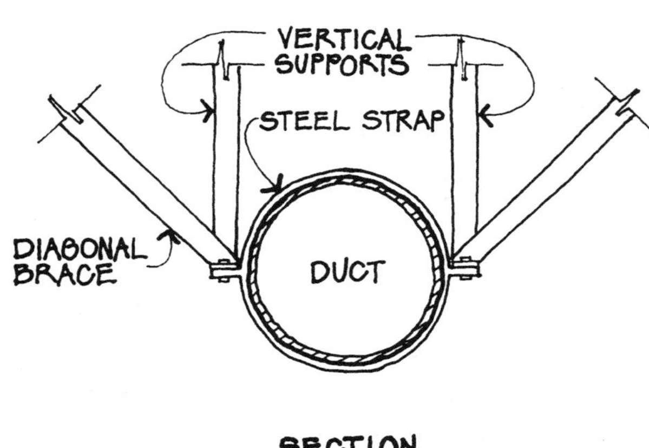

SECTION

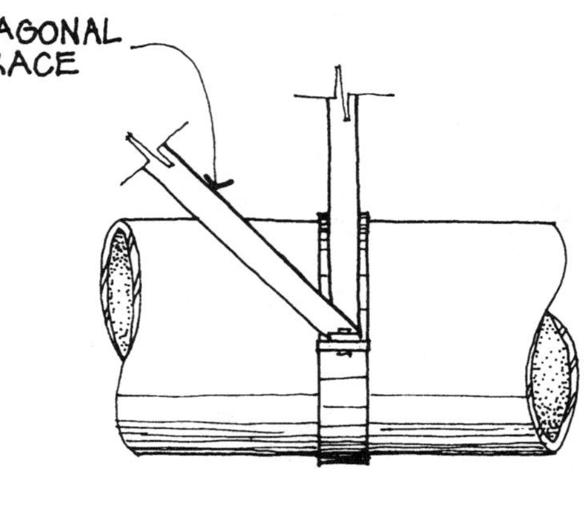

ELEVATION

TYPICAL DUCT DETAIL

9. Wood stud walls should be adequately bolted to the foundation, to prevent the structure from sliding off the foundation.
10. Wood stud cripple walls below the first floor should be braced, to prevent the structure from falling onto the foundation.
11. Wood framed roofs or floors with straight sheathing may not have adequate diaphragm strength and may require diagonal rod bracing or plywood sheathing.
12. Split level floors should be properly connected to avoid separation.
13. Concrete and masonry walls should be adequately reinforced and anchored.
14. Unreinforced masonry walls are particularly vulnerable to damage, falling, or collapse and should be strengthened and anchored, or if necessary, removed.
15. Concrete or masonry walls should be anchored to a wood diaphragm with metal ties, rather than relying on a wood ledger, which can fail by cross-grain bending and possibly cause collapse of the roof and/or falling out of the wall. Refer to the details on page 25.
16. Diagonal cracks in masonry or concrete walls may have been caused by previous earthquakes and should be analyzed and repaired.
17. Metal deck floors and roofs without a reinforced concrete topping slab may not have adequate diaphragm strength and may require strengthening.
18. The restoration of historic buildings in seismic zones poses a special challenge. The preferred, but often difficult, solution involves the reinforcement of existing elements, rather than the addition of new nonhistoric elements such as visible walls or braces.

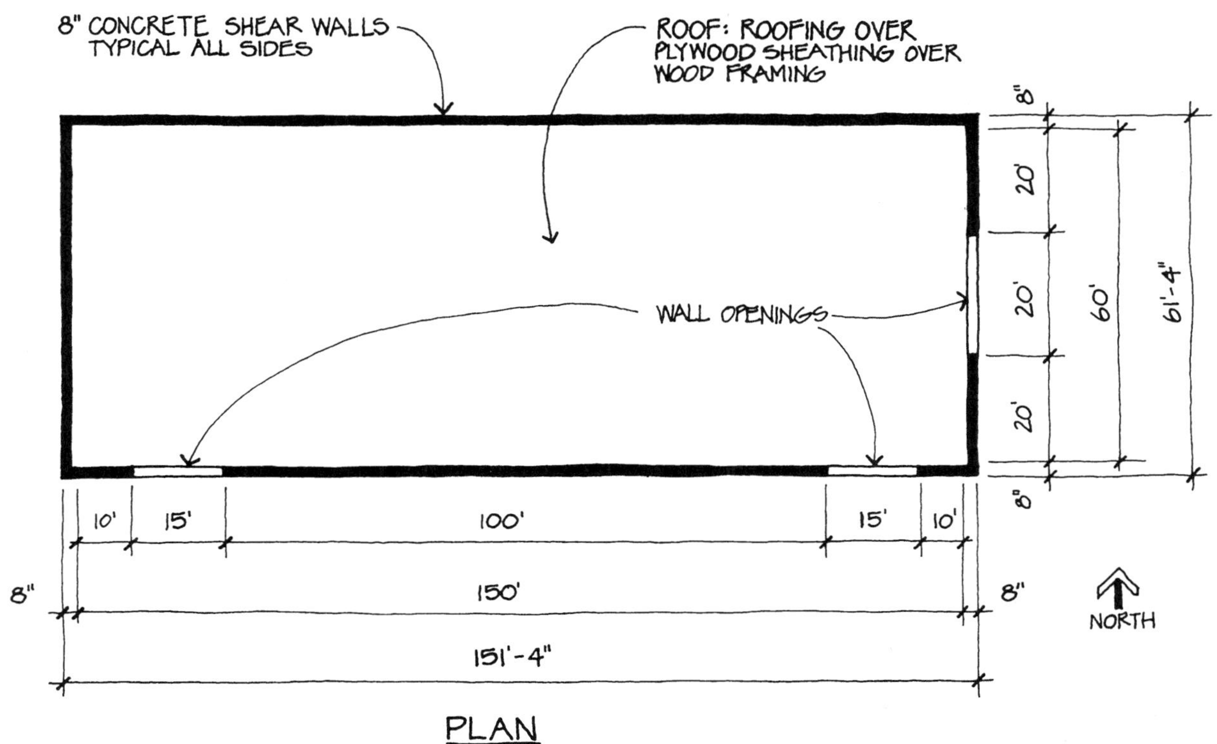

EARTHQUAKE DESIGN EXAMPLE

The following is a step-by-step design of a small commercial building in Seismic Zone 4, shown on the following two pages, to resist earthquake forces. The exam will not have a problem as lengthy or complex as this. However, if you study this design example, it should help you understand the basic principles of earthquake design.

Loads

The loads are as follows:

Roof live load	20.0 psf
Roof dead load	
Roofing	3.0
Plywood sheathing	1.5
Wood framing	2.5
Insulation	2.0
Ceiling	3.0
Mech. & elec.	2.0
Misc.	2.0
Total	16.0 psf

Dead load of 8" concrete walls
= 8"/12" × 150#/ft³ = 100.0 psf

Base Shear

What is the total base shear?

Solution:

We determine the base shear from the formula

$$V = \frac{ZIC}{R_w} W \text{ (see page 7).}$$

From Table 16-I on page 8, Z in Seismic Zone 4 = 0.40.

This building is not an essential or hazardous facility, so from Table 16-K on page 10, I = 1.0.

Since we are not given any information to the contrary, we must assume that the building does not contain a complete frame providing support for vertical loads.

Therefore, its structural system is a bearing wall system. From Table 16-N on page 16, this is system 1.2.a, which has an R_w value of 6.

From page 9, $T = C_t(h_n)^{3/4} = 0.020(14)^{3/4} = 0.145$ seconds. Note that h_n is the height to the diaphragm, not to the top of the parapet.

Since the soil properties are not known, S is assumed to be 1.5 (see page 9).

$$C = \frac{1.25S}{T^{2/3}} = \frac{1.25(1.5)}{(0.145)^{2/3}} = 6.79$$

However, the maximum value of C is 2.75 (see page 9).

$$\text{Therefore, } V = \frac{ZIC}{R_w} W = \frac{0.40 \times 1.0 \times 2.75}{6} W$$
$$= 0.183 \text{ W}$$

We next determine the dead load W as follows:

Roof dead load = 16 × 60 × 150 = 144,000#

North and south
walls dead load = 100 × 17 × 150 × 2 = 510,000#

East and west
walls dead load = 100 × 17 × 61.33 × 2 = 208,522#

Total dead load = 862,522#

V = 0.183 W = 0.183(862,522) = *157,842#*

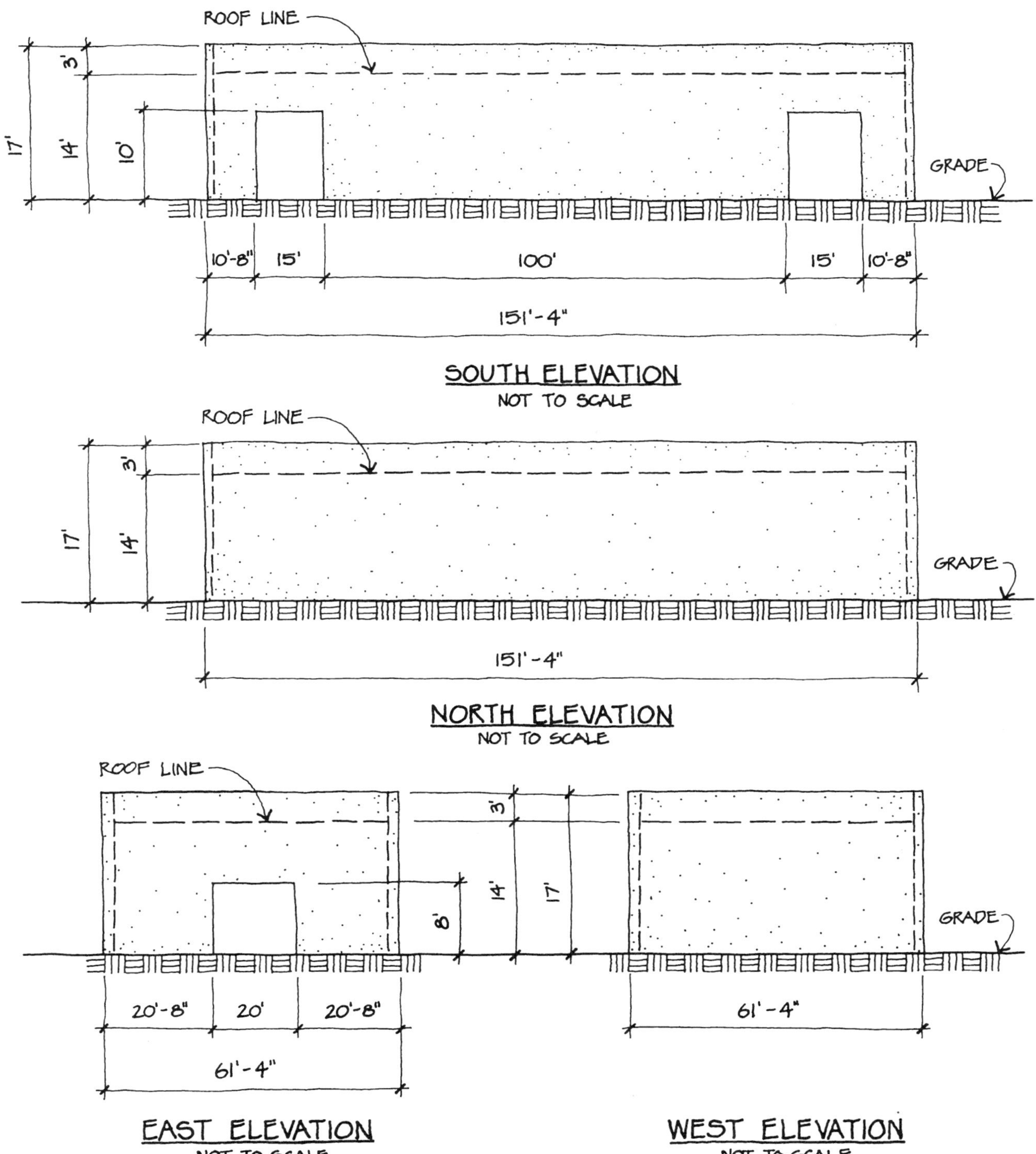
ROOF LINE
3'
14'
17'
10'
GRADE
10'-8"
15'
100'
15'
10'-8"
151'-4"
SOUTH ELEVATION
NOT TO SCALE
ROOF LINE
3'
14'
17'
GRADE
151'-4"
NORTH ELEVATION
NOT TO SCALE
ROOF LINE
3'
14'
17'
8'
GRADE
20'-8"
20'
20'-8"
61'-4"
61'-4"
EAST ELEVATION
NOT TO SCALE
WEST ELEVATION
NOT TO SCALE

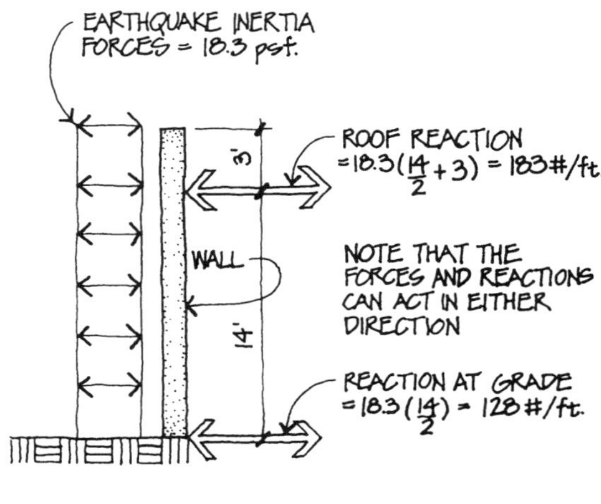

SECTION

SUPPORT OF WALL FOR LATERAL LOADS

Diaphragm Design for North-South Direction

Design the roof diaphragm. Determine the diaphragm shear, select a plywood diaphragm, and calculate the maximum chord force in the diaphragm.

Solution:

The north and south walls span vertically between grade and the roof diaphragm when subject to earthquake inertia forces. These forces are equal to 0.183 × wall weight = 0.183 × 100 = 18.3 pounds per square foot as shown above.

The bottom reaction of 128 pounds per foot is resisted by the footings and soil.

The top reaction of 183 pounds per foot is resisted by the roof diaphragm. In addition, the diaphragm resists the inertia forces caused by the roof dead load = 0.183 × 16 × 60 = 176 pounds per foot.

The roof diaphragm acts as a horizontal beam spanning between the east and west shear walls and resists a total load of 183 + 183 + 176 = 542 pounds per foot, as shown on the following page.

The plywood roof diaphragm is a flexible diaphragm, which acts as a simple beam.

The reaction to each shear wall is therefore 542#/ft. × 150 ft./2 = 542 × 75 = 40,650#.

The diaphragm shear is equal to this reaction divided by the diaphragm width = 40,650#/60 ft. = *678#/ft.*

We select a plywood diaphragm having an allowable shear of at least 678#/ft. from UBC Table 23-I-J-1, shown on page 24. Several diaphragms meet this requirement, and we select the following:

15/32" plywood, Structural I, with 10d nails at 2" o.c. at boundaries and continuous panel edges parallel to load, and at 3" o.c. at other panel edges, with blocking at all unsupported panel edges.

This diaphragm has an allowable shear of 730#/ft., which is greater than the 678#/ft. required.

The chord force is the tension or compression in the flanges, or chords, of the diaphragm. Since the diaphragm acts like a uniformly loaded simple beam, its maximum moment = $wL^2/8$.

The chord force is the maximum diaphragm moment divided by the diaphragm depth

$$= \frac{542\#/\text{ft.} \times (150\text{ ft.})^2}{8 \times 60\text{ ft.}} = 25{,}406\# = \mathit{25{,}406\#}$$

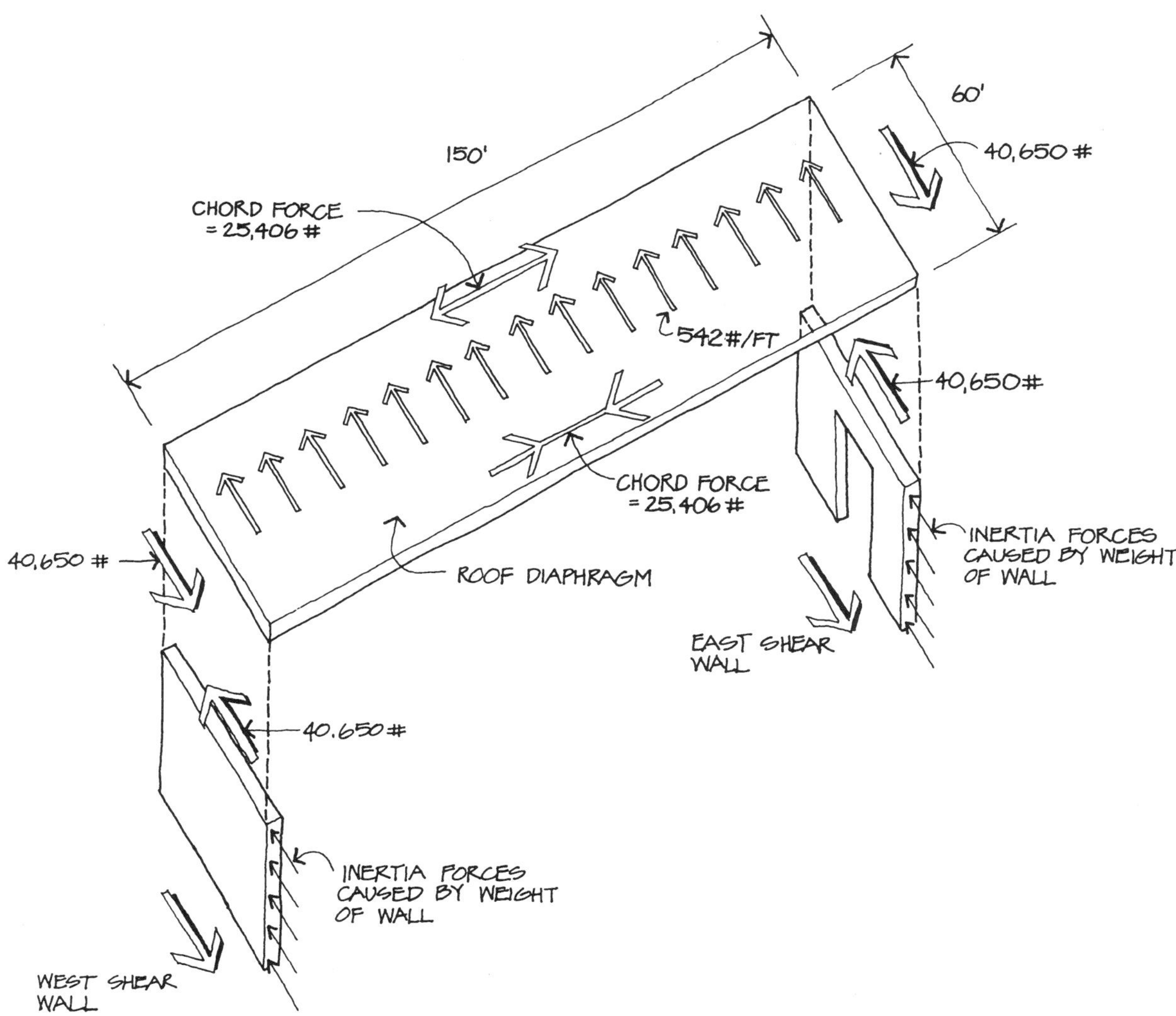

This chord force can be resisted by a continuous wood or steel member directly attached to the diaphragm, or by reinforcing steel in the north and south walls adjacent to the roof diaphragm.

Connection of Roof Diaphragm to East and West Shear Walls

The roof diaphragm is connected to the concrete walls with a wood ledger and anchor bolts, as shown in the details on page 25. What should be the maximum spacing of 3/4" anchor bolts to transfer the diaphragm shear into the walls? Neglect any vertical loads to be transferred by the anchor bolts, and assume an allowable load of 1,400# per bolt in the wood ledger and 2,940# per bolt in the concrete wall.

Solution:

The diaphragm shear travels through the plywood diaphragm and is transferred from the plywood to the ledger by the boundary nailing, and from the ledger to the concrete shear wall by the anchor bolts. The anchor bolts must be able to transfer the diaphragm shear of 678#/ft. Since the allowable bolt shear in the concrete wall is greater than its allowable shear in the wood ledger, the wood ledger value must be used in the design.

The required anchor bolt spacing is equal to

$$\frac{1{,}400\#/\text{bolt}}{678\#/\text{ft.}} = 2.06 \text{ ft/bolt} = 24.7 \text{ inches/bolt}$$

Use 3/4" anchor bolts at 24 inches on center.

East and West Shear Walls

The east and west shear walls must resist the shear from the diaphragm, which is 40,650#, plus the inertia forces caused by the weight of the walls themselves. The walls must be adequate to resist this shear. Note that the unit shear stress is greater in the east wall because the 20-foot opening reduces the net area of the wall available to resist the shear stress.

The walls must also resist the overturning moment caused by the shear. If the resisting moment from the dead load of the wall and roof is less than the overturning moment, the wall must be tied to the footing to provide additional overturning resistance.

Because of the building's configuration, earthquake forces in the north-south direction cause greater stresses in the diaphragm and shear walls than east-west forces. Therefore, in this case, design for the east-west direction is not necessary.

OTHER SEISMIC CRITERIA

As previously stated, much of the discussion in this lesson is based on the earthquake regulations in the Uniform Building Code (UBC). This is because the UBC is more widely used for earthquake design than any other model building code, and the earthquake regulations in the UBC are comprehensive and regularly updated.

However, it is possible that some exam questions might be based on other design criteria or regulations. For example, there is a two-volume publication called *NEHRP Recommended Provisions for the Development of Seismic Regulations for New Buildings*, which is published by the Building Seismic Safety Council in Washington, D.C., and listed as a reference by NCARB.

Although the seismic design concepts in this publication are generally very similar to those in the UBC, there are some differences, the most important of which are summarized below.

1. NEHRP (National Earthquake Hazards Reduction Program) uses seismic coefficients A_a and A_v which vary from 0.05 to 0.40 in accordance with detailed maps and represent effective peak accelerations. These coefficients are comparable to Z, the UBC seismic zone factor (see page 7).
2. NEHRP coefficient C_s is approximately equal to the UBC quantity $\frac{ZC}{R_w}$ (see page 9).
3. NEHRP coefficient R is comparable to UBC coefficient R_w (see page 9).
4. NEHRP Seismic Hazard Exposure Groups are comparable to UBC Occupancy Categories (see page 10).
5. The UBC includes an importance factor I, while NEHRP does not. Instead, NEHRP specifies more severe design requirements for more critical buildings and/or more earthquake-prone sites. This is accomplished by using Seismic Performance Categories which vary from A to E, A being the least severe and E the most.

UBC 1997 EDITION

The Uniform Building Code (UBC) is updated every three years, and the 1997 edition includes a number of revisions to the earthquake regulations. These changes may not necessarily be reflected on the exam, since they do not substantially alter the basic concepts of seismic design. Nevertheless, there are some important modifications, which should be recognized by exam candidates. Some of the key changes are summarized below.

1. Where strength design or load and resistance factor design (LRFD) is used, the code specifies the factored load combinations (dead, live, snow, earthquake, and wind). Alternatively, working stress load combinations may be used.
2. A redundancy/reliability factor (ρ) has been introduced, which encourages the use of redundant lateral force resisting systems.
3. Near-source factors (N_a and N_v) for Seismic Zone 4 have been added, which account for amplified ground motions where structures are close to a fault.
4. A new set of soil profile categories (S_A through S_F) has been adopted.
5. Lateral force requirements for elements of structures, nonstructural components, and equipment have been revised.
6. A simplified lateral force procedure is permitted for certain light frame structures up to three stories in height.
7. The old R_w factor has been replaced by a new R factor, which is based on strength design and is approximately equal to $R_w \div 1.4$.

These revisions are design refinements based largely on the performance of buildings in recent earthquakes, including the 1994 Northridge earthquake, as well as ongoing research.

SUMMARY

The way a building responds to an earthquake depends on many factors, including the nature of the earthquake itself, the soil type, the distance of the structure from the earthquake's epicenter, and the form, structural system, mass, structural materials, and construction quality of the building. Like a chain, a building is only as strong as its weakest element, and you can be sure that an earthquake will always discover what that is.

It is therefore important for architects to exercise sound judgment when designing structures in earthquake-prone areas. In this regard, the following is a summary of recommended seismic design practices, some of which have been covered in detail in this lesson.

1. Buildings should be as regular and symmetrical as possible, to avoid torsion and/or stress concentrations.
2. Drift limitations should be strictly adhered to, to prevent excessive movement. Adjacent buildings or portions of buildings

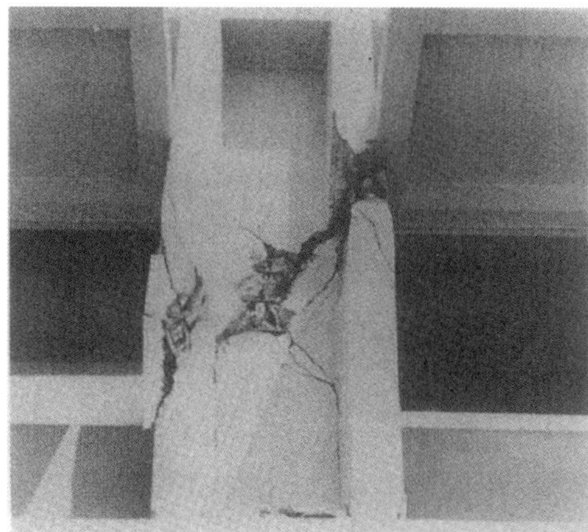

Diagonal cracks in concrete piers between windows, Olive View Hospital, San Fernando earthquake, 1971 (N.O.A.A. photograph)

should have adequate seismic separations to avoid pounding.

3. There should be a direct path for force transfer. Lateral load resisting systems should be continuous to the foundation.
4. Overturning effects should be minimized by keeping the building's height-to-width ratio low and by locating heavy elements as low in the building as possible.
5. All structural elements should be interconnected so that they move as a unit and thereby reduce the possibility of local failure.
6. All nonstructural elements should have adequate lateral bracing.
7. Redundancy should be provided by secondary systems that can resist part of the lateral force if the primary system fails or is damaged.
8. Infilling of a moment-resisting frame with a nonstructural wall tends to stiffen the frame and thereby attract more lateral load. The design and detailing of the frame and wall should therefore provide for this. Alternatively, the infilling wall should be structurally separated from the frame so that it can move independently, and yet be held in place.
9. Stairs should be designed and detailed to resist the lateral forces they may attract, or else detached from the building structure.
10. Openings in shear walls that may result in stress concentrations should be avoided, or if unavoidable, they should be adequately reinforced.

The subject of earthquake design is often mystifying or intimidating to architects, especially those from areas that seldom or never experience earthquakes. In this lesson, we hope we have made earthquake design clearer and more comprehensible, so that architectural candidates can approach the exam with confidence.

LESSON 1 QUIZ

1. The allowable shear in a blocked plywood diaphragm depends on all of the following EXCEPT

   A. plywood thickness.

   B. width of framing members.

   C. nail spacing.

   D. direction of framing members.

2. In a tall building, which of the following systems is usually most effective in resisting earthquakes?

   A. Shear walls at the perimeter.

   B. Moment-resisting frames at the perimeter.

   C. Braced frames at the perimeter.

   D. Moment-resisting frames set back from the perimeter and rigidly anchored to the perimeter elements.

3. Select the INCORRECT statement.

   A. A shear wall resists lateral forces in its own plane.

   B. A concrete diaphragm distributes horizontal forces in proportion to the rigidities of the vertical resisting elements.

   C. A plywood diaphragm distributes horizontal forces to the vertical resisting elements in proportion to the tributary area of each element.

   D. Torsion is the rotation that occurs in a diaphragm when the center of mass does not coincide with the centroid of the building.

4. Vertical ground motions are usually neglected in earthquake design because

   I. an earthquake causes only horizontal ground shaking.

   II. the vertical motions are generally smaller than the horizontal motions.

   III. most buildings have considerable excess strength in the vertical direction.

   IV. design live loads already include an earthquake factor.

   A. I only

   B. II, III, and IV

   C. II and IV

   D. II and III

5. A building in Seismic Zone 3 has a steel frame that supports all the vertical load and concrete shear walls designed to resist all the lateral earthquake load. What is the R_w value of the building?

   A. 6

   B. 8

   C. 9

   D. 12

6. A building in which core walls provide the only significant lateral force resisting elements may be inadequate to resist

   A. bending.

   B. torsion.

   C. vertical acceleration.

   D. diaphragm stresses.

7. The fundamental period of vibration of a structure

A. is unrelated to the height of the structure.

B. decreases as the height of the structure increases.

C. increases as the height of the structure increases.

D. may either increase or decrease as the height of the structure increases.

8. Special moment-resisting frames may be constructed of any of the following materials, EXCEPT

A. cast-in-place concrete.

B. precast concrete.

C. structural steel.

D. reinforced masonry.

9. Buildings designed in accordance with the Uniform Building Code earthquake provisions are believed to be

A. capable of resisting major earthquakes without damage.

B. capable of resisting shaking from major earthquakes without collapse.

C. capable of resisting ground movement, sliding, subsidence, or faulting due to an earthquake.

D. capable of resisting shaking from major earthquakes without collapse, if the lateral load resisting system is constructed of structural steel.

10. The Richter scale is a measure of an earthquake's

A. magnitude.

B. intensity.

C. acceleration.

D. force.

LESSON TWO

WIND DESIGN

NATURE OF WIND FORCES

Although windstorms are not as unpredictable as earthquakes, they are equally important in the design of buildings. In the United States, the loss of life and damage to property caused by windstorms exceeds that from earthquakes, because of their more frequent occurrence and the large geographic areas affected. Also, winds are an important design factor to a greater or lesser degree in all parts of the country, while earthquakes are predominant only in specific geographic areas.

Wind is the movement of the earth's atmosphere caused by the unequal heating of the earth's surface and the rotation of the earth. These atmospheric movements are modified by local topographical conditions, often causing violent recurrent windstorms, such as the *chinooks* of the eastern slopes of the Rockies, and the *Santa Anas* in Southern California.

Severe tropical storms are called *hurricanes* and occur mainly along the Gulf and South Atlantic coasts, although they have extended as far north as New England. Hurricane wind speeds usually vary between 30 and 120 miles per hour, with gusts up to 180 mph. Winds of 50 to 60 mph can strip small branches and leaves off trees. At 70 to 80 mph, shallow rooted trees and thin walls can be blown down, an entire roof sometimes lifted off, and large glass windows blown in. Above 80 mph, lifting of roofs and snapping of trees are common.

Tornadoes are the most violent of atmospheric storms, and consist of rotating air masses of small diameter. They are much smaller than hurricanes, but have greater wind speeds. The rotational speed of a tornado is estimated to be nearly 250 miles per hour, and may occasionally exceed 500 mph. The paths of tornadoes average several hundred yards in width and 16 miles in length, although large variations from these averages are not uncommon.

MEASUREMENT OF WINDS

Wind speed is measured by a device called a *revolving cup anemometer*. In the design of buildings, we are concerned with the maximum average wind speed, called the *fastest mile speed*. This is determined by the time it takes a column of air one mile long to pass the anemometer, the mile being measured by a specific number of revolutions of the anemometer. This measurement thus averages out the gusts and lulls, making wind measurements comparable from one measuring station to another.

The wind speed also varies with the height above the ground, since the friction between the wind and the ground surface reduces the velocity of the wind. To make the wind station measurements comparable, all wind speed data have been standardized to a height of 10 meters, or 33 feet.

These fastest mile speeds, or basic wind speeds, are shown in Figure 16-1 of the Uniform Building Code, reproduced below by permission of the International Conference of Building Officials (IBC Figure 1609 and Table 1609.3.1, see

FIGURE 16-1—MINIMUM BASIC WIND SPEEDS IN MILES PER HOUR (× 1.61 FOR KM/H)

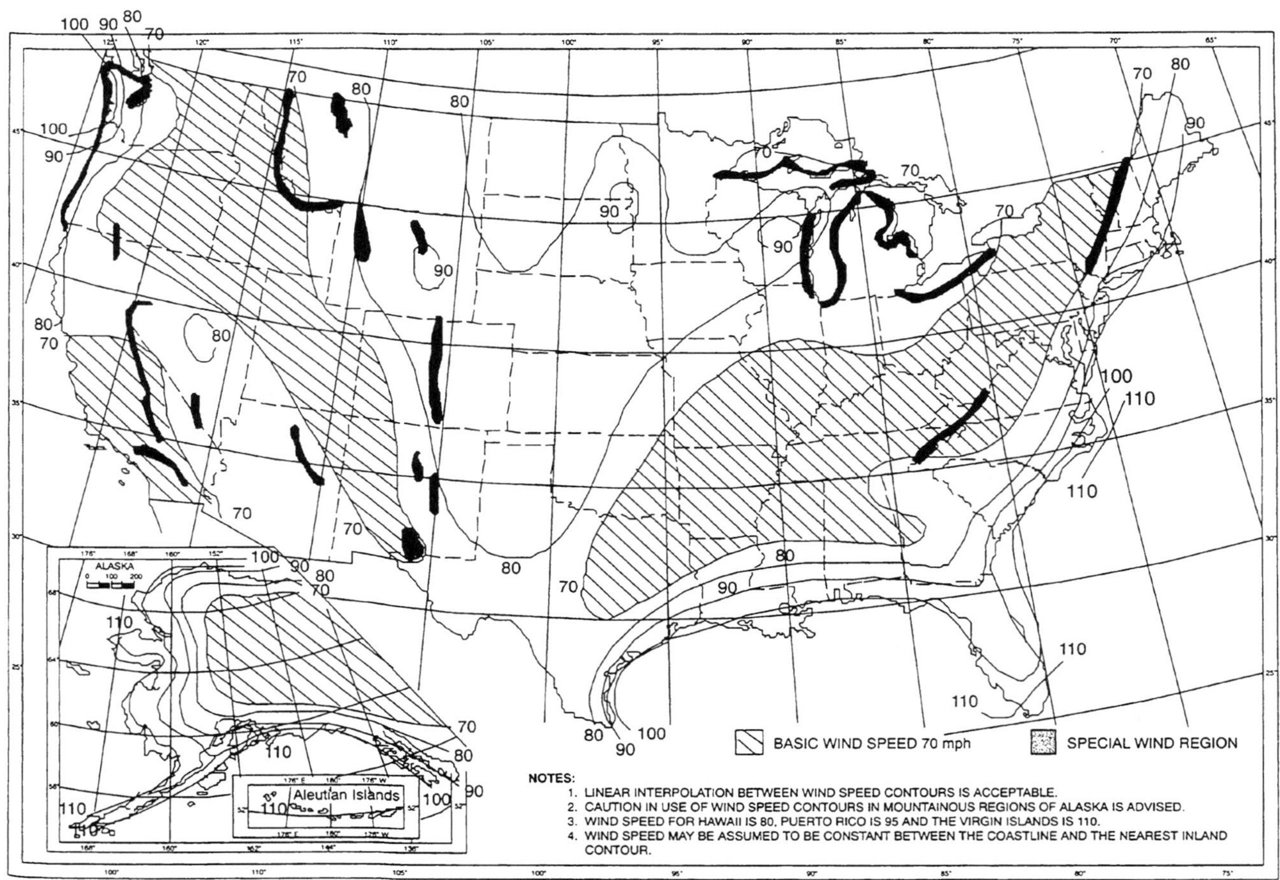

Appendix). The wind speed contours on the map were determined from a limited number of stations. Accordingly, when using the map, great attention must be paid to local wind records and conditions, which may indicate higher basic wind speeds. In fact, areas shown on the map as *special wind regions*, which include the Pacific Northwest coast and Puget Sound, the mountains and canyons of Southern California, and the shores of the Great Lakes, have experienced basic wind speeds so different from those of the surrounding geographic areas that they are specifically excluded from the map. The wind speeds in these areas must be determined from local records.

The wind records have been adjusted statistically to provide a mean recurrence interval of 50 years. This means that the fastest mile wind speed has a 2 percent probability of occurring in any one year, and should therefore occur on the average once every 50 years. As can be seen, the lowest wind speed used on the map is 70 miles per hour. Note that the areas along the Gulf and South Atlantic coasts have very high wind speeds, to account for hurricanes. Tornadoes, however, are *not* accounted for on the wind speed map.

RESPONSE OF BUILDINGS TO WIND LOADS

Buildings respond to wind forces in a variety of complex and dynamic ways. The most obvious of these is the direct pressure of the wind as it is stopped by the windward side of the building, that is, the side facing the direction from which the wind is blowing.

The wind creates a negative pressure, or *suction*, on the leeward side of the building, which is the side opposite the windward side, and the side walls parallel to the wind direction. Uplift pressure is created on horizontal or sloping surfaces, such as the roof.

The building structure also vibrates because of the gusting effects of the wind, similar to the way a building responds to earthquake motion. These vibrations are important not only because of their effect on the building structure, but also because such motions in taller, more flexible, buildings can be uncomfortable to the occupants.

As the wind passes over the building, friction causes drag forces to act in the direction of the wind.

The corners, edges, and eave overhangs of the building are subjected to especially complex forces as the wind passes these obstructions, causing much higher local suction forces than those on the building as a whole. In fact, most wind damage to buildings occurs in these areas.

These different responses of a building to wind loads are shown in the sketch on the following page. Of course, any combination of these responses, or all of them, can occur at the same time, thus making the overall building response very complex.

The direct wind pressure, also called the *stagnation pressure*, in pounds per square foot, on a vertical surface is related to the fastest mile velocity in miles per hour, by the formula $p = 0.00256\ V^2$. Thus, if the fastest mile velocity is 70 miles per hour, the wind pressure is $p = 0.00256 \times (70)^2 = 12.5$ psf. Since the pressure varies as the square of the velocity, *doubling the wind velocity increases the wind pressure fourfold*. The wind velocities and pressures for some common climatological conditions are shown in the table on the following page.

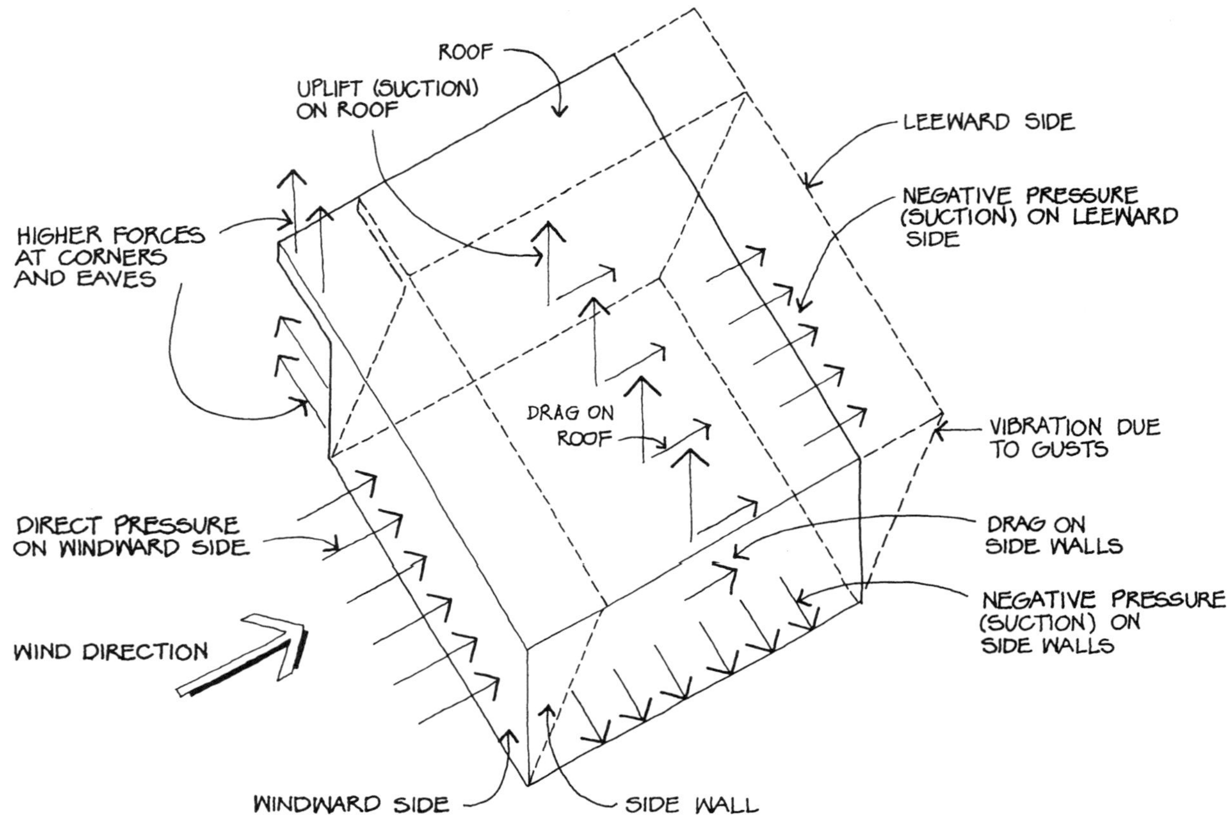

BUILDING RESPONSE TO WIND

Wind Pressure lbs per sq. ft.	Wind Velocity miles per hour	Description
160	250	Tornado
58	150	Strong Hurricane
21	90	Hurricane
9	60	Perimeter of Windstorm
2	30	Strong Breeze

There are other factors that affect the wind pressure on a building. As previously discussed, the friction between the wind and the ground causes the wind velocity, and thus the pressure, to decrease close to the ground, and conversely, to increase as the height increases.

As mentioned, wind gusts have velocities, and thus pressures, greater than the average, and these gusts must be considered in the building design.

Another important factor affecting wind velocity and pressure is the terrain surrounding the building. If it is built-up with buildings or other features, the resulting turbulence reduces the wind velocity and pressure. On the other hand, buildings on open sites, such as those on ridges or facing large bodies of water, must bear the full brunt of the wind and thus have greater pressures.

Also affecting the wind pressure are the size and shape of the building. Narrow buildings allow the wind to flow around them and therefore have lower pressures, while wide, tall buildings block more of the wind and thus have greater pressures.

CODE REQUIREMENTS

Our discussion is based on the requirements of the 1994 edition of the Uniform Building Code (UBC). Tables 16-F, 16-G, and 16-H of the UBC are reproduced in this lesson by permission of the International Conference of Building Officials.

Buildings of unusual shape, such as domes, or with unusual site conditions, such as at the mouth of a canyon, are not covered by the UBC requirements. In such situations, wind tunnel testing may have to be done to determine the applicable design loads.

The effects of tornadoes are also not considered in these code requirements, and are beyond the scope of this course.

As discussed earlier, the actual wind forces are dynamic, constantly varying in magnitude and direction. The UBC, however, uses a static analysis method, similar to that used in earthquake design, in which static or non-varying external forces are applied to the building structure that simulate the actual varying wind forces.

Basic Formula for Static Analysis

The UBC procedure consists of determining the design wind pressure on the building (p) as the product of four factors: C_e, C_q, q_s, and I.

As in earthquake design, the wind force is generally evaluated in the two horizontal directions parallel to the main axes of the building. The structure must be designed to resist the forces caused by either wind or earthquake, whichever is greater, in each direction, but not in both directions acting at the same time.

The horizontal forces determined from the basic formula represent the direct horizontal pressure against the windward side, or the suction on the leeward side, combined with the drag on the horizontal and sidewall surfaces, and the internal pressure caused by the wall openings. The uplift forces on the roof combine the external suction and the internal pressure.

The horizontal and uplift forces are assumed to act simultaneously.

C_e Factor

The factor C_e accounts for the height of the building, the exposure, or roughness of the terrain, at the building site, and an increase to account for the gusting effect. The factor is determined from Table 16-G, reproduced on page 52.

Since the value of C_e varies with the height above the average adjoining ground level, the wind pressure on the building varies with the height, as shown in the sketch on the following page.

The site exposure is classified as Exposure B, C, or D. Exposure B *has terrain with buildings, forest or surface irregularities, covering at least 20 percent of the ground level area extending one mile or more from the site*, as in most built-up areas. Exposure C is defined as having *terrain that is flat and generally open, extending one-half mile or more from the site in any full quadrant.* Exposure D *represents the most severe exposure in areas with basic wind speeds of 80 miles per hour (mph) or greater and has terrain that is flat and unobstructed facing large bodies of water over one mile or more in width relative to any quadrant of the building site. Exposure D extends inland*

TABLE 16-G — COMBINED HEIGHT, EXPOSURE, AND GUST FACTOR COEFFICIENT (C_e)[1]

HEIGHT ABOVE AVERAGE LEVEL OF ADJOINING GROUND (feet) × 304.8 for mm	EXPOSURE D	EXPOSURE C	EXPOSURE B
0-15	1.39	1.06	0.62
20	1.45	1.13	0.67
25	1.50	1.19	0.72
30	1.54	1.23	0.76
40	1.62	1.31	0.84
60	1.73	1.43	0.95
80	1.81	1.53	1.04
100	1.88	1.61	1.13
120	1.93	1.67	1.20
160	2.02	1.79	1.31
200	2.10	1.87	1.42
300	2.23	2.05	1.63
400	2.34	2.19	1.80

[1]Values for intermediate heights above 15 feet (4572 mm) may be interpolated.

Reprinted from *Uniform Building Code,* copyright International Conference of Building Officials, Whittier, California, 1994.

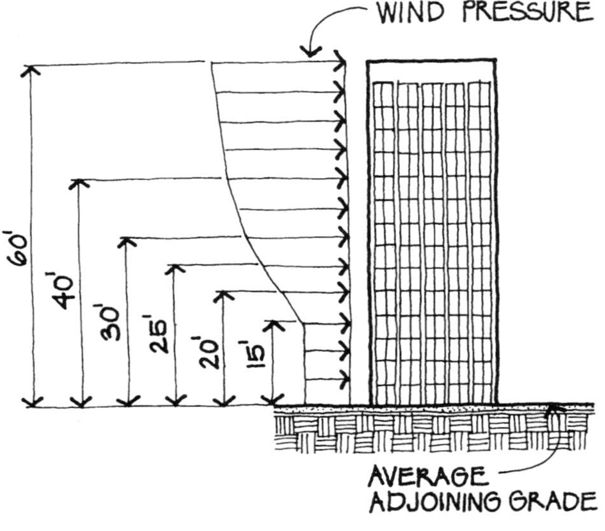

WIND PRESSURE VARIES WITH HEIGHT

from the shoreline 1/4 mile or 10 times the building height, whichever is greater. As can be seen from Table 16-G, the factors for the different exposures vary widely.

C_q Factor

The C_q factor is a pressure coefficient for the structure or portion of the structure under consideration. For example, the direct pressure on a windward wall is not the same as the suction or uplift pressure on the roof. The factor is determined from Table 16-H, a portion of which is reproduced on the following page.

Methods 1 and 2

There are two methods for determining the wind pressures to be used in the design of the building's primary lateral load resisting system.

In Method 1, called the *Normal Force Method,* the wind pressures normal (perpendicular) to all external surfaces are determined and are assumed to act simultaneously. Method 1 may be used for any building, and must be used for gabled rigid frames. For windward walls, the pressure varies with height in accordance with Table 16-G. For inward or outward pressures on roofs and leeward walls, the factor C_e is evaluated at the mean roof height and therefore these pressures are constant.

In Method 2, called the *Projected Area Method,* the wind pressures acting on the full projected horizontal and vertical areas of the building are determined and are assumed to act simultaneously. The use of Method 2 is limited to buildings less

TABLE 16-H — PRESSURE COEFFICIENTS (C_q)

STRUCTURE OR PART THEREOF	DESCRIPTION	C_q FACTOR
1. Primary frames and systems	**Method 1** (Normal force method)	
	Walls:	
	Windward wall	0.8 inward
	Leeward wall	0.5 outward
	Roofs[1]:	
	Wind perpendicular to ridge	
	Leeward roof or flat roof	0.7 outward
	Windward roof	
	less than 2:12 (16.7%)	0.7 outward
	Slope 2:12 (16.7%) to less than 9:12 (75%)	0.9 outward or 0.3 inward
	Slope 9:12 (75%) to 12:12 (100%)	0.4 inward
	Slope > 12:12 (100%)	0.7 inward
	Wind parallel to ridge and flat roofs	0.7 outward
	Method 2 (Projected area method)	
	On vertical projected area	
	Structures 40 feet (12 192 mm) or less in height	1.3 horizontal any direction
	Structures over 40 feet (12 192 mm) in height	1.4 horizontal any direction
	On horizontal projected area[1]	0.7 upward

than 200 feet in height, except those using gabled rigid frames.

In both methods, the C_e factor includes the effects of external pressure or suction, internal pressure or suction, and wind drag on the surfaces parallel to the wind direction.

The sketches below show the wind pressures on a building using Methods 1 and 2.

q_s Factor

The q_s factor is the wind stagnation pressure or direct wind pressure at a standard height of 33 feet, determined from the basic wind speed, or fastest

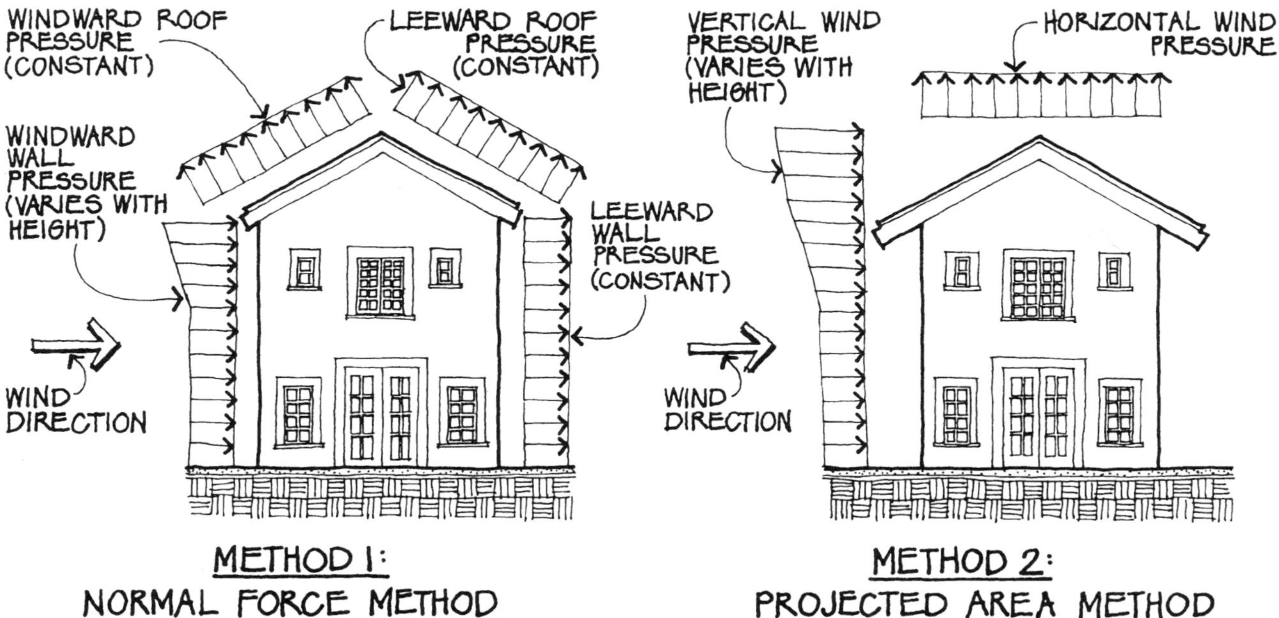

TABLE 16-F—WIND STAGNATION PRESSURE (q_s) AT STANDARD HEIGHT OF 33 FEET

Basic wind speed (mph)[1] (× 1.61 for km/h)	70	80	90	100	110	120	130
Pressure q_s (psf) (× 0.0479 for kN/m^2)	12.6	16.4	20.8	25.6	31.0	36.9	43.3

[1]Wind speed from Section 1615.

mile speed, at that height. As was previously discussed, q_s can be calculated from the formula $q_s = 0.00256\ V^2$. However, the code presents the results of this calculation in Table 16-F, shown above. Accordingly, to determine the value of q_s, one must first determine the basic wind speed at the site under consideration from UBC Figure 16-1, from local wind speed records if they indicate a higher basic speed, or from local records in the *special wind regions* indicated on the map. Then, this basic wind speed in mph is converted to pressure q_s in psf using Table 16-F.

I Factor

The factor I is the importance factor and is similar to that used in earthquake design. Referring to Table 16-K on page 10, you can see that the value of I is either 1.0 or 1.15, depending on the occupancy category of the building.

Hence, essential facilities such as hospitals and fire and police stations are designed for wind forces 15 percent greater than normal (I = 1.15). In this way, such emergency facilities are expected to be safe and usable following a severe windstorm.

Similarly, hazardous facilities must also be designed for these increased wind forces, in order to provide a greater measure of safety for such facilities.

Using an I factor of 1.15 for essential or hazardous facilities has the effect of designing such buildings for a windstorm with a mean recurrence interval of 100 years. All other buildings are designed using an I factor of 1.0, which is based on an average recurrence of once in 50 years.

Example #1

What is the basic wind speed in Salt Lake City?

Solution:

The exam sometimes tests candidates' familiarity with the wind speed map, which is Figure 16-1 in the UBC, reproduced on page 48. Using that map, we see that the basic wind speed in Salt Lake City is *70 mph.*

Example #2

What is the static horizontal and vertical wind pressure in pounds per square foot on a flat roof fire station which is 15 feet high and located in the downtown area of Los Angeles, using Method 1?

Solution:

The static horizontal wind pressure is determined from the basic UBC formula $p = C_eC_qq_sI$. From Table 16-G, shown on page 52, we read in the column for *Exposure B*, for built-up areas, that $C_e = 0.62$ for heights from 0 to 15 feet.

From Table 16-H, on page 53, we find that for Method 1 the values of C_q are 0.8 inward for the windward wall, 0.5 outward for the leeward wall, and 0.7 outward for the flat roof.

From UBC Figure 16-1, on page 48, we find that Los Angeles has a basic wind speed of 70 miles per hour, and from Table 16-F, page 54, we see that a basic wind speed of 70 miles per hour corresponds to a value of q_s of 12.6 psf.

Since this building is a fire station, the importance factor I is equal to 1.15.

Then using the basic formula $p = C_eC_qq_sI$, we find the following pressures:

Windward wall:
$p = 0.62 \times 0.8 \times 12.6 \times 1.15 =$ *7.2 psf inward*
Leeward wall:
$p = 0.62 \times 0.5 \times 12.6 \times 1.15 =$ *4.5 psf outward*
Roof:
$p = 0.62 \times 0.7 \times 12.6 \times 1.15 =$ *6.3 psf outward*

The wind pressures are shown in the sketch below.

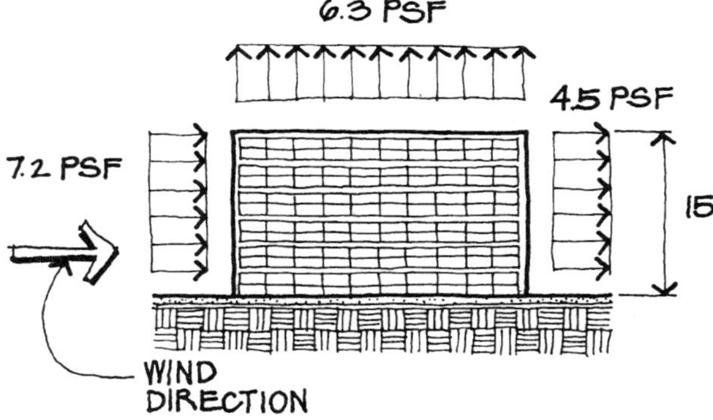

Example #3

If the building in Example #2 were located in an isolated desert region outside of Los Angeles, not in a *special wind region*, what would the wind pressures be?

Solution:

The only change from Example #2 is the site exposure, which is Exposure C, instead of the Exposure B of Example #2. From Table 16-G, the factor C_e becomes 1.06 instead of 0.62. The wind pressures would then be:

Windward wall:
$p = 1.06 \times 0.8 \times 12.6 \times 1.15 =$ *12.3 psf inward*
Leeward wall:
$p = 1.06 \times 0.5 \times 12.6 \times 1.15 =$ *7.7 psf outward*
Roof:
$p = 1.06 \times 0.7 \times 12.6 \times 1.15 =$ *10.8 psf outward*

If we compare these results with those of Example #2, we can see how much greater the wind pressures are for buildings on open exposed sites.

Example #4

If the building in Examples #2 and #3 is 50 feet by 50 feet in plan, and one story high, what is the total wind load resisted by the roof diaphragm?

Solution:

As in earthquake design, the walls are assumed to span vertically from the ground to the roof. Thus, one-half of the wind load is transferred from the walls into the roof diaphragm, which in turn transfers the load into the vertical shear resisting elements. The other half of the wind load is resisted at the bottom of the wall, at the ground. Thus we have:

For the urban site, wind load =

$$(7.2 \text{ psf} + 4.5 \text{ psf}) \times \frac{15 \text{ feet}}{2} \times 50 \text{ feet}$$

= *4,388 pounds*

For the open site, wind load =

$$(12.3 \text{ psf} + 7.7 \text{ psf}) \times \frac{15 \text{ feet}}{2} \times 50 \text{ feet}$$

= *7,500 pounds*

Example #5

For the same building as in Example #2, what would the wind pressures be if Method 2 were used instead of Method 1?

Solution:

If we use Method 2 instead of Method 1, we use a different C_q factor to determine the wind pressure. For the given building, we find from Table 16-H (page 53) that the value of C_q is 1.3 in any direction for the vertical projected area and 0.7 upward for the horizontal projected area. The factors C_e, q_s, and I are all the same as in Example #2. Then using the basic formula $p = C_e C_q q_s I$, we find the following pressures:

Vertical projection:
$p = 0.62 \times 1.3 \times 12.6 \times 1.15 =$ *11.7 psf in any horizontal direction*

Horizontal projection:
$p = 0.62 \times 0.7 \times 12.6 \times 1.15 =$ *6.3 psf upward*

These pressures are shown in the sketch above right.

Example #6

Rework Example #3 using Method 2.

Solution:

Vertical projection:
$p = 1.06 \times 1.3 \times 12.6 \times 1.15 =$ *20.0 psf in any horizontal direction*

Horizontal projection:
$p = 1.06 \times 0.7 \times 12.6 \times 1.15 =$ *10.8 psf upward*

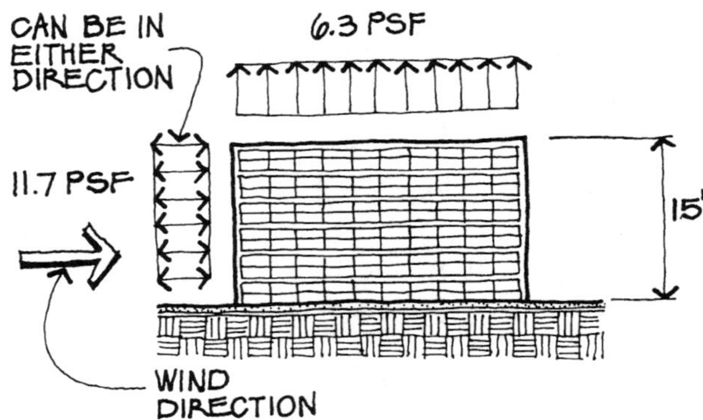

Example #7

Rework Example #4 using Method 2.

Solution:

Using the wind pressures from Examples #5 and #6, we find for the urban site, wind load =

$$11.7 \text{ psf} \times \frac{15 \text{ feet}}{2} \times 50 \text{ feet}$$

= *4,388 pounds*

For the open site, wind load =

$$20.0 \text{ psf} \times \frac{15 \text{ feet}}{2} \times 50 \text{ feet}$$

= *7,500 pounds*

Comparing these results with those of Example 4, we find that for this small building, the wind loads calculated using Methods 1 and 2 are identical.

LATERAL LOAD RESISTING SYSTEMS

The lateral load resisting systems used in wind design are the same as those used in earthquake design, namely moment-resisting frames, shear walls, and braced frames.

Unlike earthquake design, however, the type of lateral load resisting system used does not affect the magnitude of the wind load. Also, the concept of ductility, or the ability of a system to absorb energy in the inelastic range, is less important for wind design than for earthquake design. The reason is that in earthquake design, the system is expected to be stressed in the inelastic range, above the yield point, while the stresses in wind design are expected to be in the elastic range, below the yield point.

Except for these differences, the general descriptions of moment-resisting frames, shear walls, and braced frames found in Lesson 1 for earthquake design also apply to wind design.

OVERTURNING

As with earthquake forces, buildings must be designed to resist the overturning moments caused by wind forces. Because wind forces act primarily in one direction for a longer period of time than earthquake forces, which constantly reverse direction, the dead load resisting moment must be at least 1-1/2 times the wind overturning moment.

Put another way, the wind overturning moment may not exceed 2/3 of the dead load resisting moment, unless the structure is anchored to resist the excess moment.

Example #8

A two-story braced frame is subject to the wind loads and dead loads shown above right. What is the overturning moment at the base? What is the dead load resisting moment? Does the structure have an adequate factor of safety against overturning?

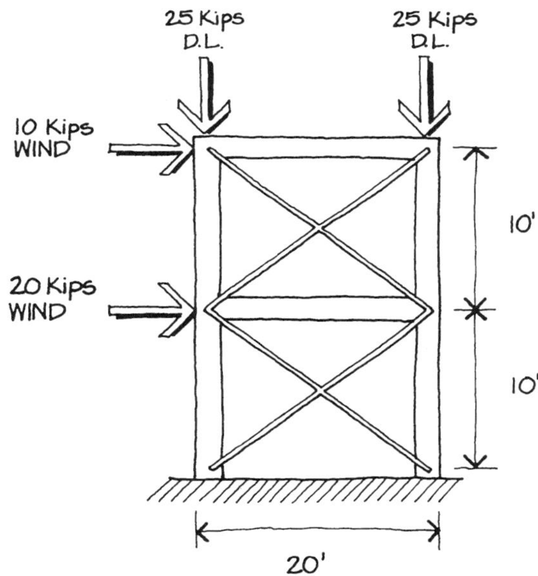

Solution:

The overturning moment = 10 kips (10 + 10) ft. + 20 kips (10 ft.) = 200 + 200 = *400 ft.-kips.*

The dead load resisting moment = 25 kips (20 ft.) = *500 ft.-kips.*

The factor of safety against overturning = dead load resisting moment ÷ overturning moment = 500 ft.-kips ÷ 400 ft.-kips = 1.25.

Since this is less than 1.5, *the structure does not have an adequate factor of safety against overturning.*

If the columns are adequately anchored to the foundation, the weight of the foundation may be used to increase the dead load resisting moment.

As with earthquake design, buildings with a high height-to-width ratio are most critical for overturning from wind forces. Thus, tall, slender buildings tend to overturn more than short, squat buildings, and top-heavy buildings tend to overturn more than pyramidal buildings.

DEFLECTION AND DRIFT

The deflection or drift of a building under wind loading must be limited in order to prevent damage to the brittle elements of the building and to minimize discomfort to the building's occupants. The drift between adjacent stories is generally limited to 0.0025 times the story height, which is about half of that usually allowed for earthquake forces, because high wind loads tend to occur more often than earthquakes

As with earthquake design, care must be taken to adequately separate adjacent buildings, and nonstructural elements should be detailed to allow for movement under wind loads.

DIAPHRAGMS, COLLECTORS, AND TORSION

Wind forces are transferred to the vertical resisting elements (shear walls, braced frames, or moment-resisting frames) by diaphragms, in exactly the same way as earthquake forces. Therefore, the discussion of diaphragms in Lesson 1 for earthquake design also applies to wind design.

Likewise, collector members may be used in wind design, just as in earthquake design, and the discussion in Lesson 1 on this subject also applies to wind design.

As described in Lesson 1 for earthquake design, torsion occurs in a rigid diaphragm when the center of mass does not coincide with the center of rigidity. Similarly, under wind loading, torsion occurs in a rigid diaphragm when the center of the applied wind load does not coincide with the center of rigidity. The discussion of torsion in Lesson 1 is therefore applicable to wind design, except that accidental torsion is not generally considered in wind design.

ELEMENTS AND COMPONENTS OF STRUCTURES

The wind pressures on elements and components of a building are determined from the same basic formula as that used for the design of the primary lateral force resisting system: $p = C_e C_q\ q_s\ I$.

However, at any given instant, the wind pressures over the surface of the building vary greatly, and gusts cause the pressures in localized areas to exceed the average pressure on the entire building.

Furthermore, observed wind damage to buildings and wind tunnel tests indicate that very high suctions or negative pressures occur at the building's discontinuities, such as eaves, ridges, and wall corners.

Therefore, higher values for the C_q factor are used in the design of elements, components, and discontinuities. These factors are shown in the portion of Table 16-H shown on page 59. The other factors in the pressure formula (C_e, q_s, and I) remain the same as for the design of the primary lateral force resisting system.

Example #9

If the building in Example #2 has a four-foot-high parapet wall, with the top at 15 feet above the adjoining ground, for how much moment should the wall be designed?

Solution:

$$p = C_e\ C_q q_s\ I$$

From Table 16-H on the following page, we find that $C_q = 1.3$ inward or outward for parapets.

TABLE 16-H — PRESSURE COEFFICIENT (C_q)

STRUCTURE OR PART THEREOF	DESCRIPTION	C_q FACTOR
2. Elements and components not in areas of discontinuity[2]	Wall elements All structures Enclosed and unenclosed structures Partially enclosed structures Parapets walls	 1.2 inward 1.2 outward 1.6 outward 1.3 inward or outward
	Roof elements[3] Enclosed and unenclosed structures Slope < 7:12 (58.3%) Slope 7:12 (58.3%) to 12:12 (100%) Partially enclosed structures Slope < 2:12 (16.7%) Slope 2:12 (16.7%) to 7:12 (58.3%) Slope > 7:12 (58.3%) to 12:12 (100%)	 1.3 outward 1.3 outward or inward 1.7 outward 1.6 outward or 0.8 inward 1.7 outward or inward
3. Elements and components in areas of discontinuities[2,4,5]	Wall corners[6] Roof eaves, rakes or ridges without overhangs[6] Slope < 2:12 (16.7%) Slope 2:12 (16.7%) to 7:12 (58.3%) Slope > 7:12 (58.3%) to 12:12 (100%) For slopes less than 2:12 (16.7%) Overhangs at roof eaves, rakes or ridges, and canopies	1.5 outward or 1.2 inward 2.3 upward 2.6 outward 1.6 outward 0.5 added to values above

From Example #2, $C_e = 0.62$, $q_s = 12.6$ psf and I = 1.15. Therefore, p = 0.62 × 1.3 × 12.6 × 1.15 = 11.7 psf inward or outward.

The moment = 11.7 psf × 4 ft. × 4ft./2 = *93.6'#*

Example #10

For the parapet wall in Example #9, what is the moment at the corner of the building?

Solution:

From Table 16-H, we find that $C_q = 1.5$ outward or 1.2 inward for wall corners. Therefore, p = 0.62 × 1.5 × 12.6 × 1.15 = 13.5 psf.

The moment = 13.5 psf × 4 ft. × 4 ft./2 = *108'#*

COMBINED VERTICAL AND HORIZONTAL FORCES

The UBC requires that all building components be designed for the following load combinations:

1. Dead plus floor live plus wind.
2. Dead plus floor live plus wind + snow/2.
3. Dead plus floor live plus snow + wind/2.

Note that the roof live load and seismic load are not considered to act concurrently with the wind load, since the likelihood is very remote that they will be acting on the structure during a windstorm.

Also, only one-half of the wind or snow load is used in combinations 2 and 3 because it is very unlikely that both these loads will have their maximum values at the same time.

Allowable stresses may be increased one-third when considering wind forces either acting alone or combined with the vertical loads above.

WIND DESIGN EXAMPLE

The following is a step-by-step design of a small commercial building built to resist wind forces. The building is identical to that used in the earthquake design example that starts on page 38 and is assumed to be located in an urban area where the basic wind speed is 70 mph.

As with the earthquake design example, the exam will not have a problem as lengthy or complex as this. However, careful study of this design example should help you understand the basic principles of wind design.

Wind Pressure

What is the wind pressure on the vertical projected area using Method 2?

Solution:

Determine the wind pressure using the basic UBC formula $p = C_e C_q q_s I$

Since this building is in an urban area, its exposure is Exposure B, and from Table 16-G on page 52, C_e = 0.62 up to 15 feet and 0.67 at 20 feet. By interpolation, C_e at 17 feet = 0.62 + 2/5(0.67 − 0.62) = 0.62 + .02 = 0.64.

From Table 16-H, shown on page 53, using Method 2, C_q = 1.3 horizontal on the vertical projected area.

From Table 16-F on page 54, for a basic wind speed of 70 mph, q_s = 12.6 psf.

This building is not an essential or hazardous facility, so I = 1.0.

We now have all of the factors in the wind pressure equation, and

p (up to 15 feet) = 0.62 × 1.3 × 12.6 psf × 1.0 = *10.2 psf*

p (at 17 feet) = 0.64 × 1.3 × 12.6 psf × 1.0 = *10.5 psf*

Diaphragm Design for North-South Direction

Design the roof diaphragm. Determine the diaphragm shear, select a plywood diaphragm, and calculate the maximum chord force in the diaphragm.

Solution:

The north and south walls span vertically between grade and the roof diaphragm when subject to wind load, as shown below.

The bottom reaction of 71 pounds per foot is resisted by the footings and soil.

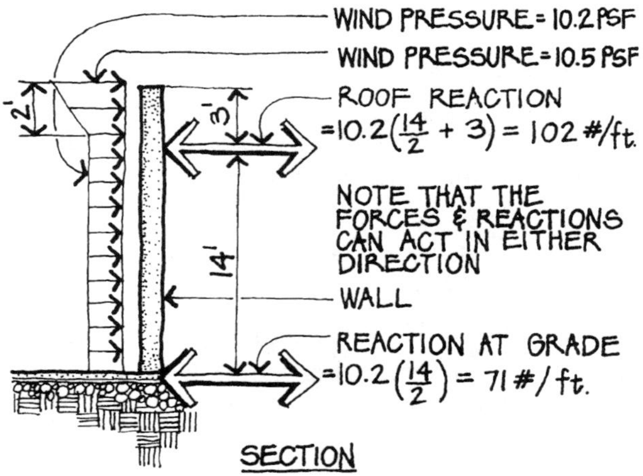

SUPPORT OF WALL FOR LATERAL LOADS

The top reaction of 102 pounds per foot is resisted by the roof diaphragm, which acts as a horizontal beam spanning between the east and west shear walls, as shown on page 62. Note that in calculating the value of the top reaction, as shown below, we assumed the wind pressure to have a constant value of 10.2 psf and neglected the slightly higher pressure at the top of the wall.

The plywood roof diaphragm is a flexible diaphragm, which acts as a simple beam. The reaction to each shear wall is therefore 102#/ft. × 150 ft./2 = 7,650#.

The diaphragm shear is equal to this reaction divided by the diaphragm width = 7,650#/60 ft. = *128#/ft.*

We select a plywood diaphragm having an allowable shear of at least 128#/ft. from UBC Table No. 23-I-J-1, shown on page 24. Several diaphragms meet this requirement, and we select the following: *3/8" plywood, C-C, with 8d nails at 6" o.c. at supported panel edges, without blocking at the panel edges.* This diaphragm has an allowable shear of 160#/ft., which is greater than the 128#/ft. required.

The chord force is the tension or compression in the flanges, or chords, of the diaphragm. Since the diaphragm acts like a uniformly loaded simple beam, its maximum moment = $wL^2/8$.

The chord force is the maximum diaphragm moment divided by the diaphragm depth

$$= \frac{102\#/\text{ft.} \times (150\text{ft.})^2}{8 \times 60 \text{ ft.}} = \mathit{4{,}781\#}$$

As discussed in the earthquake design example, this chord force can be resisted by a continuous wood or steel member, or by reinforcing steel in the north and south walls.

Connection of Roof Diaphragm to East and West Shear Walls

Refer to the earthquake design example on page 38. What should be the maximum spacing of the anchor bolts to transfer the diaphragm shear into the walls?

Solution:

The anchor bolts must be able to transfer the diaphragm shear of 128#/ft.

The required anchor bolt spacing is equal to

$$\frac{1{,}400\#/\text{bolt}}{128\#/\text{ft.}} = 10.9 \text{ ft./bolt}$$

While a bolt spacing of 10 feet on center would be adequate, it is common practice to space anchor bolts at no more than *four feet on center.*

East and West Shear Walls

The east and west shear walls must be adequate to resist the shear from the diaphragm, which is 7,650#. Note that the unit shear stress is greater in the east wall because the 20-foot opening reduces the net area of the wall available to resist the shear stress.

The walls must also resist the overturning moment caused by the shear. If the resisting moment from the dead load of the wall and roof is less than 1.5 times the overturning moment, the wall must be tied to the footing to provide additional overturning resistance.

Because of the building's configuration, wind forces in the north-south direction cause greater stresses in the diaphragm and shear walls than

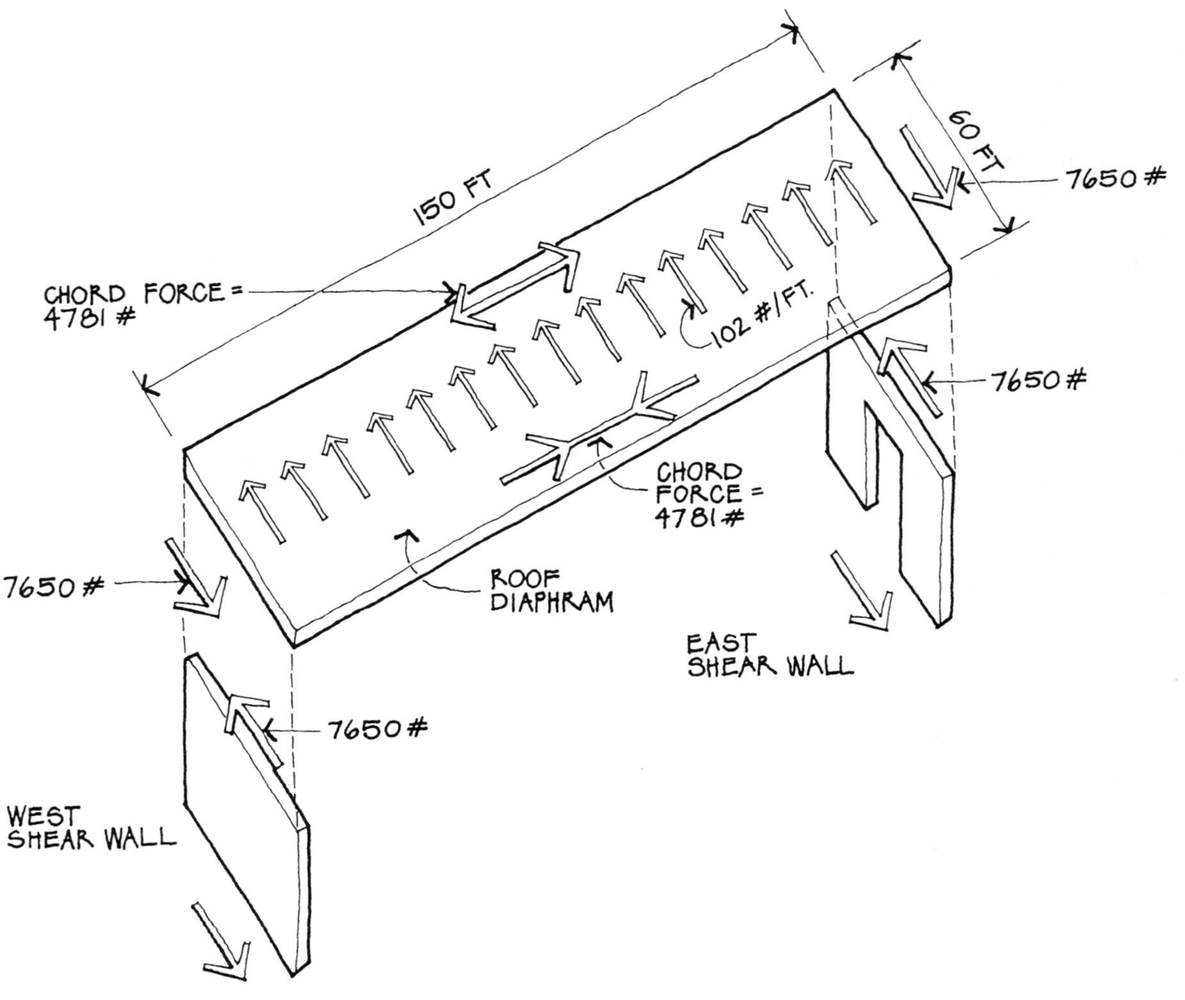

east-west forces. Therefore, in this case, design for the east-west direction is not necessary.

Comparing this wind design with the earthquake design for the same building in Seismic Zone 4, we can see that the earthquake stresses are greater than the wind stresses and therefore govern the design of the building. In actual practice, once having determined that the wind load to the roof diaphragm is much less than the seismic load to the diaphragm, it would not be necessary to proceed with the remainder of the wind design; the members and connections designed to resist the earthquake forces would be more than adequate to resist the wind forces.

CONCLUSION

Wind design is very much like earthquake design. For both, we want to minimize torsion and overturning effects, avoid building shapes which could result in stress concentrations, adhere strictly to drift limitations, and provide redundancy.

In this context, a number of recommended practices for earthquake design are discussed in Lesson 1 (see page 38). These generally apply to wind design as well, though to a somewhat lesser degree, and should therefore be reviewed.

LESSON 2 QUIZ

1. Using Method 1, the wind pressure on a building is assumed to act

   I. inward on the windward wall.

   II. outward on the leeward wall.

   III. upward on a flat roof.

   IV. inward on the leeward wall.

   V. downward on a flat roof.

   VI. outward on the windward wall.

   A. I, II, and III **C.** II, III, and VI

   B. I, III, and IV **D.** I, II, and V

2. When designing buildings for wind forces, which of the following are NOT considered?

   I. Drift

   II. Liquefaction

   III. Tsunamis

   IV. Overturning moment

   A. I only **C.** I, II, and III

   B. II and III **D.** I, II, III, and IV

3. A building 15 feet high with Exposure C and located in an area with a basic wind speed of 70 mph should be designed for what pressure on the windward wall, using Method 1? Assume I = 1.0.

   A. 6.2 psf **C.** 6.7 psf

   B. 10.7 psf **D.** 14.0 psf

4. The Uniform Building Code wind load coefficients account for which of the following?

   I. Height of the building

   II. Exposure of the site

   III. Hurricanes and tornadoes

   IV. Wind gusting

   A. I only **C.** I, II, and IV

   B. II and III **D.** I, II, III, and IV

5.

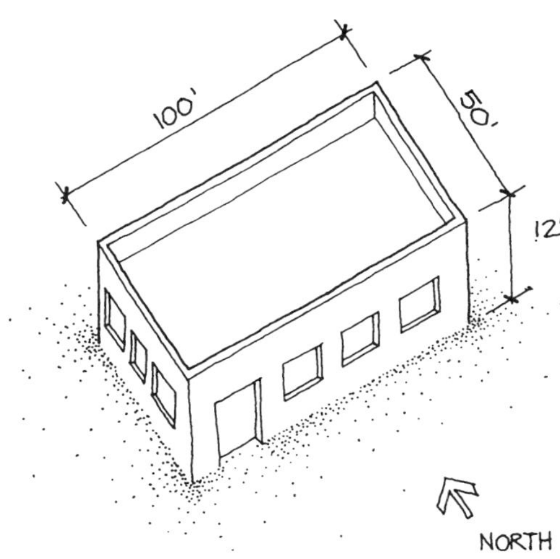

Using the projected area method, the one-story building shown above has a design wind pressure (p) on the vertical projected area of 20 psf. For wind acting in the north-south direction, what is the total wind force to be transferred from the roof diaphragm to the east or west shear wall?

   A. 3 kips **C.** 12 kips

   B. 6 kips **D.** 24 kips

6. For the building in the preceding question, what is the maximum diaphragm shear?

   A. 120#/ft.
   B. 240#/ft.
   C. 60#/ft.
   D. 480#/ft.

7. Elements and components of structures are designed for wind pressures that are

   A. less than those used for the design of the primary lateral force resisting system.
   B. the same as those used for the design of the primary lateral force resisting system.
   C. greater than those used for the design of the primary lateral force resisting system.
   D. unrelated to those used for the design of the primary lateral force resisting system.

8. A shear wall resisting wind forces is subject to an overturning moment of 400 ft.-kips. The dead load resisting moment is 500 ft.-kips. Which of the following statements is correct?

   A. The shear wall has an adequate factor of safety against overturning.
   B. The shear wall does not have an adequate factor of safety against overturning, and must therefore be made longer and/or heavier.
   C. The shear wall does not have an adequate factor of safety against overturning, and must therefore have sufficient anchorage to the foundation to prevent overturning.
   D. None of the above statements are correct, since buildings are not usually designed to resist the overturning moments caused by wind forces.

9. Which of the following methods is likely to be most economical when designing a very tall building to resist wind forces?

   A. Moment-resisting frames
   B. Shear walls
   C. Base isolation
   D. Tubular system

10. All of the following load combinations must be considered in the design of building components, EXCEPT

   A. dead plus roof live plus wind.
   B. dead plus floor live plus wind.
   C. dead plus floor live plus wind plus snow/2.
   D. dead plus floor live plus snow plus wind/2.

APPENDIX

This appendix contains several tables used or referenced by the International Building Code. As discussed earlier in the "Purpose and Procedure" section of this book, the ARE only requires you to familiarize yourself with any one of the codes in use today. Any exam question involving codes will provide you with the table or equation needed to answer it correctly.

The tables and figures that appear in this appendix are used courtesy of the International Code Council. For information about the International Code Council (ICC) codes and related services, please contact ICC customer service at 1-800-786-4452.

TABLE 1604.5—CLASSIFICATION OF BUILDINGS AND OTHER STRUCTURES FOR IMPORTANCE FACTORS

CATEGORY[a]	NATURE OF OCCUPANCY	SEISMIC FACTOR I_E	SNOW FACTOR I_S	WIND FACTOR I_W
I	Buildings and other structures that represent a low hazard to human life in the event of failure including, but not limited to: • Agricultural facilities • Certain temporary facilities • Minor storage facilities	1.00	0.8	0.87[b]
II	Buildings and other structures except those listed in Categories I, III and IV	1.00	1.0	1.00
III	Buildings and other structures that represent a substantial hazard to human life in the event of failure including, but not limited to: • Buildings and other structures where more than 300 people congregate in one area • Buildings and other structures with elementary school, secondary school or day care facilities with an occupant load greater than 250 • Buildings and other structures with an occupant load greater than 500 for colleges or adult education facilities • Health care facilities with an occupant load of 50 or more resident patients but not having surgery or emergency treatment facilities • Jails and detention facilities • Any other occupancy with an occupant load greater than 5,000 • Power-generating stations, water treatment for potable water, waste water treatment facilities and other public utility facilities not included in Category IV • Buildings and other structures not included in Category IV containing sufficient quantities of toxic or explosive substances to be dangerous to the public if released	1.25	1.1	1.15
IV	Buildings and other structures designed as essential facilities including, but not limited to: • Hospitals and other health care facilities having surgery or emergency treatment facilities • Fire, rescue and police stations and emergency vehicle garages • Designated earthquake, hurricane or other emergency shelters • Designated emergency preparedness, communication, and operation centers and other facilities required for emergency response • Power-generating stations and other public utility facilities required as emergency backup facilities for Category IV structures • Structures containing highly toxic materials as defined by Section 307 where the quantity of the material exceeds the maximum allowable quantities of Table 307.7(2) • Aviation control towers, air traffic control centers and emergency aircraft hangars • Buildings and other structures having critical national defense functions • Water treatment facilities required to maintain water pressure for fire suppression	1.50	1.2	1.15

a. For the purpose of Section 1616.2, Categories I and II are considered Seismic Use Group I, Category III is considered Seismic Use Group II and Category IV is equivalent to Seismic Use Group III.

b. In hurricane-prone regions with $V > 100$ miles per hour, I_w shall be 0.77.

TABLE 2306.4.1 — ALLOWABLE SHEAR (POUNDS PER FOOT) FOR WOOD STRUCTURAL PANEL SHEAR WALLS WITH FRAMING OF DOUGLAS-FIR-LARCH, OR SOUTHERN PINE[a] FOR WIND OR SEISMIC LOADING[b,h,i,j]

TABLE 2306.4.1
ALLOWABLE SHEAR (POUNDS PER FOOT) FOR WOOD STRUCTURAL PANEL SHEAR WALLS WITH FRAMING OF DOUGLAS-FIR-LARCH, OR SOUTHERN PINE[a] FOR WIND OR SEISMIC LOADING[b, h, i, j]

PANEL GRADE	MINIMUM NOMINAL PANEL THICKNESS (inch)	MINIMUM FASTENER PENETRATION IN FRAMING (inches)	PANELS APPLIED DIRECT TO FRAMING					PANELS APPLIED OVER 1/2" OR 5/8" GYPSUM SHEATHING				
			NAIL (common or galvanized box) or staple size[k]	Fastener spacing at panel edges (inches)				NAIL (common or galvanized box) or staple size[k]	Fastener spacing at panel edges (inches)			
				6	4	3	2[e]		6	4	3	2[e]
Structural I Sheathing	5/16	1 1/4	6d	200	300	390	510	8d	200	300	390	510
		1	1 1/2 16 Gage	165	245	325	415	2 16 Gage	125	185	245	315
	3/8	1 3/8	8d	230[d]	360[d]	460[d]	610[d]	10d	280	430	550[f]	730
		1	1 1/2 16 Gage	155	235	315	400	2 16 Gage	155	235	310	400
	7/16	1 3/8	8d	255[d]	395[d]	505[d]	670[d]	10d	280	430	550[f]	730
		1	1 1/2 16 Gage	170	260	345	440	2 16 Gage	155	235	310	400
	15/32	1 3/8	8d	280	430	550	730	10d	280	430	550[f]	730
		1	1 1/2 16 Gage	185	280	375	475	2 16 Gage	155	235	300	400
		1 1/2	10d	340	510	665[f]	870	10d	—	—	—	—
Sheathing, plywood siding[g] except Group 5 Species	5/16 or 1/4[c]	1 1/4	6d	180	270	350	450	8d	180	270	350	450
		1	1 1/2 16 Gage	145	220	295	375	2 16 Gage	110	165	220	285
	3/8	1 1/4	6d	200	300	390	510	8d	200	300	390	510
		1 3/8	8d	220[d]	320[d]	410[d]	530[d]	10d	260	380	490[f]	640
		1	1 1/2 16 Gage	140	210	280	360	2 16 Gage	140	210	280	360
	7/16	1 3/8	8d	240[d]	350[d]	450[d]	585[d]	10d	260	380	490[f]	640
		1	1 1/2 16 Gage	155	230	310	395	2 16 Gage	140	210	280	360
	15/32	1 3/8	8d	260	380	490	640	10d	260	380	490[f]	640
		1 1/2	10d	310	460	600[f]	770	—	—	—	—	—
		1	1 1/2 16 Gage	170	255	335	430	2 16 Gage	140	210	280	360
	19/32	1 1/2	10d	340	510	665[f]	870	—	—	—	—	—
		1	1 3/4 16 Gage	185	280	375	475	—	—	—	—	—
			Nail Size (galvanized casing)					Nail Size (galvanized casing)				
	5/16[c]	1 1/4	6d	140	210	275	360	8d	140	210	275	360
	3/8	1 3/8	8d	160	240	310	410	10d	160	240	310[f]	410

For SI: 1 inch = 25.4 mm, 1 pound per foot = 14.5939 N/m.

a. For framing of other species: (1) Find specific gravity for species of lumber in AF&PA National Design Specification. (2) For staples find shear value from table above for Structural I panels (regardless of actual grade) and multiply value by 0.82 for species with specific gravity of 0.42 or greater, or 0.65 for all other species. (3) For nails find shear value from table above for nail size for actual grade and multiply value by the following adjustment factor: Specific Gravity Adjustment Factor = [1-(0.5 - SG)], where SG = Specific Gravity of the framing lumber. This adjustment factor shall not be greater than 1.

b. Panel edges backed with 2-inch nominal or wider framing. Install panels either horizontally or vertically. Space fasteners maximum 6 inches on center along intermediate framing members for 3/8-inch and 7/16-inch panels installed on studs spaced 24 inches on center. For other conditions and panel thickness, space fasteners maximum 12 inches on center on intermediate supports.

c. 3/8-inch panel thickness or siding with a span rating of 16 inches on center is the minimum recommended where applied direct to framing as exterior siding.

d. Shears are permitted to be increased to values shown for 15/32-inch sheathing with same nailing provided (a) studs are spaced a maximum of 16 inches on center, or (b) if panels are applied with long dimension across studs.

e. Framing at adjoining panel edges shall be 3 inches nominal or wider, and nails shall be staggered where nails are spaced 2 inches on center.

f. Framing at adjoining panel edges shall be 3 inches nominal or wider, and nails shall be staggered where both of the following conditions are met: (1) 10d nails having penetration into framing of more than 1 1/2 inches and (2) nails are spaced 3 inches on center.

g. Values apply to all-veneer plywood. Thickness at point of fastening on panel edges governs shear values.

h. Where panels are applied on both faces of a wall and nail spacing is less than 6 inches o.c. on either side, panel joints shall be offset to fall on different framing members. Or framing shall be 3 inch nominal or thicker and nails on each side shall be staggered.

i. In Seismic Design Category D, E or F, where shear design values exceed 490 pounds per lineal foot (LRFD) or 350 pounds per lineal foot (ASD) all framing members receiving edge nailing from abutting panels shall not be less than a single 3-inch nominal member. Plywood joint and sill plate nailing shall be staggered in all cases. See Section 2305.3.10 for sill plate side and anchorage requirements.

j. Galvanized nails shall be hot dipped or tumbled.

k. Staples shall have a minimum crown width of 7/16 inch.

TABLE 1617.6.2—DESIGN COEFFICIENTS AND FACTORS FOR BASIC SEISMIC-FORCE-RESISTING SYSTEMS

TABLE 1617.6.2—continued
DESIGN COEFFICIENTS AND FACTORS FOR BASIC SEISMIC-FORCE-RESISTING SYSTEMS

BASIC SEISMIC-FORCE-RESISTING SYSTEM	DETAILING REFERENCE SECTION	RESPONSE MODIFICATION COEFFICIENT, R^a	SYSTEM OVERSTRENGTH FACTOR, Ω_0^g	DEFLECTION AMPLIFICATION FACTOR, C_d^b	SYSTEM LIMITATIONS AND BUILDING HEIGHT LIMITATIONS (FEET) BY SEISMIC DESIGN CATEGORY[c] AS DETERMINED IN SECTION 1616.3[e]				
					A or B	C	D[d]	E[e]	F[e]
H. Ordinary plain concrete shear walls	1910.2.1	2	2½	2	NL	NP	NP	NP	NP
I. Composite eccentrically braced frames	(14)[k]	8	2	4	NL	NL	160	160	100
J. Composite concentrically braced frames	(13)[k]	5	2	4½	NL	NL	160	160	100
K. Ordinary composite braced frames	(12)[k]	3	2	3	NL	NL	NP	NP	NP
L. Composite steel plate shear walls	(17)[k]	6½	2½	5½	NL	NL	160	160	100
M. Special composite reinforced concrete shear walls with steel elements	(16)[k]	6	2½	5	NL	NL	160	160	100
N. Ordinary composite reinforced concrete shear walls with steel elements	(15)[k]	5	2½	4½	NL	NL	NP	NP	NP
O. Special reinforced masonry shear walls	1.13.2.2.5[o]	5½	2½	4	NL	NL	160	160	100
P. Intermediate reinforced masonry shear walls	1.13.2.2.4[o]	4	2½	2½	NL	NL	NP	NP	NP
Q. Ordinary reinforced masonry shear walls	1.13.2.2.3[o]	3	2½	2¼	NL	160	NP	NP	NP
R. Detailed plain masonry shear walls	1.13.2.2.2[o]	2½	2½	2¼	NL	NP	NP	NP	NP
S. Ordinary plain masonry shear walls	1.13.2.2.1[o]	1½	2½	1¼	NL	NP	NP	NP	NP
T. Light frame walls with shear panels—wood structural panels/sheet steel panels	2306.4.1/ 2211	7	2½	4½	NL	NL	65	65	65
U. Light framed walls with shear panels—all other materials	2306.4.5/ 2211	2½	2½	2½	NL	NL	35	NP	NP
V. Ordinary plain prestressed masonry shear walls	2106.1.1.1	1½	2½	1¼	NL	NP	NP	NP	NP
W. Intermediate prestressed masonry shear walls	2106.1.1.2, 1.13.2.2.4[o]	3	2½	2½	NL	35	NP	NP	NP
X. Special prestressed masonry shear walls	2106.1.1.3, 1.13.2.2.5[o]	4½	2½	4	NL	35	35	35	35
3. Moment-resisting Frame Systems									
A. Special steel moment frames	(9)[j]	8	3	5½	NL	NL	NL	NL	NL
B. Special steel truss moment frames	(12)[j]	7	3	5½	NL	NL	160	100	NP
C. Intermediate steel moment frames	(10)[j]	4½	3	4	NL	NL	35[h]	NP[h,i]	NP[h,i]
D. Ordinary steel moment frames	(11)[j]	3½	3	3	NL	NL	NP[h,i]	NP[h,i]	NP[h,i]
E. Special reinforced concrete moment frames	(21.1)[l]	8	3	5½	NL	NL	NL	NL	NL

TABLE 1617.6.2 (continued)—DESIGN COEFFICIENTS AND FACTORS FOR BASIC SEISMIC-FORCE-RESISTING SYSTEMS

TABLE 1617.6.2—continued
DESIGN COEFFICIENTS AND FACTORS FOR BASIC SEISMIC-FORCE-RESISTING SYSTEMS

BASIC SEISMIC-FORCE-RESISTING SYSTEM	DETAILING REFERENCE SECTION	RESPONSE MODIFICATION COEFFICIENT, R^a	SYSTEM OVERSTRENGTH FACTOR, Ω_0^g	DEFLECTION AMPLIFICATION FACTOR, C_d^b	SYSTEM LIMITATIONS AND BUILDING HEIGHT LIMITATIONS (FEET) BY SEISMIC DESIGN CATEGORY AS DETERMINED IN SECTION 1616.3[c]				
					A or B	C	D[d]	E[e]	F[e]
F. Intermediate reinforced concrete moment frames	(21.1)[l]	5	3	4 1/2	NL	NL	NP	NP	NP
G. Ordinary reinforced concrete moment frames	(21.1)[l]	3	3	2 1/2	NL	NP	NP	NP	NP
H. Special composite moment frames	(9)[k]	8	3	5 1/2	NL	NL	NL	NL	NL
I. Intermediate composite moment frames	(10)[k]	5	3	4 1/2	NL	NL	NP	NP	NP
J. Composite partially restrained moment frames	(8)[k]	6	3	5 1/2	160	160	100	NP	NP
K. Ordinary composite moment frames	(11)[k]	3	3	2 1/2	NL	NP	NP	NP	NP
L. Masonry wall frames	2106	5 1/2	3	5	NL	NL	160	160	100
4. Dual Systems with Special Moment Frames									
A. Steel eccentrically braced frames, moment-resisting connections, at columns away from links	(15)[j]	8	2 1/2	4	NL	NL	NL	NL	NL
B. Steel eccentrically braced frames, nonmoment-resisting connections, at columns away from links	(15)[j]	7	2 1/2	4	NL	NL	NL	NL	NL
C. Special steel concentrically braced frames	(13)[j]	8	2 1/2	6 1/2	NL	NL	NL	NL	NL
D. Special reinforced concrete shear walls	1910.2.4	8	2 1/2	6 1/2	NL	NL	NL	NL	NL
E. Ordinary reinforced concrete shear walls	1910.2.3	7	2 1/2	6	NL	NL	NP	NP	NP
F. Composite eccentrically braced frames	(14)[k]	8	2 1/2	4	NL	NL	NL	NL	NL
G. Composite concentrically braced frames	(13)[k]	6	2 1/2	5	NL	NL	NL	NL	NL
H. Composite steel plate shear walls	(17)[k]	8	2 1/2	6 1/2	NL	NL	NL	NL	NL
I. Special composite reinforced concrete shear walls with steel elements	(16)[k]	8	2 1/2	6 1/2	NL	NL	NL	NL	NL
J. Ordinary composite reinforced concrete shear walls with steel elements	(15)[k]	7	2 1/2	6	NL	NL	NP	NP	NP
K. Special reinforced masonry shear walls	1.13.2.2.5[o]	7	3	6 1/2	NL	NL	NL	NL	NL
L. Intermediate reinforced masonry shear walls	1.13.2.2.4[o]	6 1/2	3	5 1/2	NL	NL	NP	NP	NP
5. Dual Systems with Intermediate Moment Frames[m]									
A. Special steel concentrically braced frames[f]	(13)[j]	4 1/2	2 1/2	4	NL	NL	35[h]	NP[h,i]	NP
B. Special reinforced concrete shear walls	1910.2.4	6	2 1/2	5	NL	NL	160	100	100
C. Ordinary reinforced concrete shear walls	1910.2.3	5 1/2	2 1/2	4 1/2	NL	NL	NP	NP	NP

TABLE 1617.6.2 (continued)—DESIGN COEFFICIENTS AND FACTORS FOR BASIC SEISMIC-FORCE-RESISTING SYSTEMS

TABLE 1617.6.2–continued
DESIGN COEFFICIENTS AND FACTORS FOR BASIC SEISMIC-FORCE-RESISTING SYSTEMS

BASIC SEISMIC-FORCE-RESISTING SYSTEM	DETAILING REFERENCE SECTION	RESPONSE MODIFICATION COEFFICIENT, R^a	SYSTEM OVERSTRENGTH FACTOR, Ω_0^g	DEFLECTION AMPLIFICATION FACTOR, C_d^b	SYSTEM LIMITATIONS AND BUILDING HEIGHT LIMITATIONS (FEET) BY SEISMIC DESIGN CATEGORY AS DETERMINED IN SECTION 1616.3[c]				
					A or B	C	D[d]	E[e]	F[e]
D. Ordinary reinforced masonry shear walls	1.13.2.2.3[o]	3	3	$2^1/_2$	NL	160	NP	NP	NP
E. Intermediate reinforced masonry shear walls	1.13.2.2.4[o]	5	3	$4^1/_2$	NL	NL	NP	NP	NP
F. Composite concentrically braced frames	(13)[k]	5	$2^1/_2$	$4^1/_2$	NL	NL	160	100	NP
G. Ordinary composite braced frames	(12)[k]	4	$2^1/_2$	3	NL	NL	NP	NP	NP
H. Ordinary composite reinforced concrete shear walls with steel elements	(15)[k]	$5^1/_2$	$2^1/_2$	$4^1/_2$	NL	NL	NP	NP	NP
6. Shear Wall-frame Interactive System with Ordinary Reinforced Concrete Moment Frames and Ordinary Reinforced Concrete Shear Walls	21.1[l] 1910.2.3	$5^1/_2$	$2^1/_2$	5	NL	NP	NP	NP	NP
7. Inverted Pendulum Systems									
A. Cantilevered column systems		$2^1/_2$	2	$2^1/_2$	NL	NL	35	35	35
B. Special steel moment frames	(9)[j]	$2^1/_2$	2	$2^1/_2$	NL	NL	NL	NL	NL
C. Ordinary steel moment frames	(11)[j]	$1^1/_4$	2	$2^1/_2$	NL	NL	NP	NP	NP
D. Special reinforced concrete moment frames	21.1[l]	$2^1/_2$	2	$1^1/_4$	NL	NL	NL	NL	NL
8. Structural Steel Systems Not Specifically Detailed for Seismic Resistance	AISC—335 AISC—LRFD AISI AISC—HSS	3	3	3	NL	NL	NP	NP	NP

For SI: 1 foot = 304.8 mm, 1 pound per square foot = 0.0479 KN/m^2.

a. Response modification coefficient, R, for use throughout.

b. Deflection amplification factor, C_d.

c. NL = Not limited and NP = Not permitted.

d. See Section 1617.6.2.4.1 for a description of building systems limited to buildings with a height of 240 feet or less.

e. See Section 1617.6.2.4.1 for building systems limited to buildings with a height of 160 feet or less.

f. Ordinary moment frame is permitted to be used in lieu of intermediate moment frame in Seismic Design Categories B and C.

g. The tabulated value of the overstrength factor, Ω_o, is permitted to be reduced by subtracting $^1/_2$ for structures with flexible diaphragms but shall not be taken as less than 2.0 for any structure.

h. Steel ordinary moment frames and intermediate moment frames are permitted in single-story buildings up to a height of 60 feet, when the moment joints of field connections are constructed of bolted end plates and the dead load of the roof does not exceed 15 pounds per square foot. The dead weight of the portion of walls more than 35 feet above the base shall not exceed 15 pounds per square foot.

i. Steel ordinary moment frames are permitted in buildings up to a height of 35 feet, where the dead load of the walls, floors and roof does not exceed 15 pounds per square foot.

j. AISC 341 Part I or Part III section number.

k. AISC 341 Part II section number.

l. ACI 318, Section number.

m. Steel intermediate moment resisting frames as part of a dual system are not permitted in Seismic Design Categories D, E, and F.

n. Steel ordinary concentrically braced frames are permitted in penthouse structures and in single-story buildings up to a height of 60 feet when the dead load of the roof does not exceed 15 pounds per square foot.

o. ACI 530/ASCE 5/TMS 402 section number.

TABLE 1616.5.1—PLAN STRUCTURAL IRREGULARITIES

TABLE 1616.5.1.1 PLAN STRUCTURAL IRREGULARITIES

	IRREGULARITY TYPE AND DESCRIPTION	REFERENCE SECTION	SEISMIC DESIGN CATEGORY[a] APPLICATION
1a	Torsional Irregularity—to be considered when diaphragms are not flexible as determined in Section 1602.1.1 Torsional irregularity shall be considered to exist when the maximum story drift, computed including accidental torsion, at one end of the structure transverse to an axis is more than 1.2 times the average of the story drifts at the two ends of the structure.	9.5.5.5.2 of ASCE 7 1620.4.1 9.5.2.5.1 of ASCE 7 9.5.5.7.1 of ASCE 7	C, D, E and F D, E and F D, E and F C, D, E and F
1b	Extreme Torsional Irregularity—to be considered when diaphragms are not flexible as determined in Section 1602.1. Extreme torsional irregularity shall be considered to exist when the maximum story drift, computed and including accidental torsion, at one end of the structure transverse to an axis is more than 1.4 times the average of the story drifts at the two ends of the structure.	9.5.5.5.2 of ASCE 7 1620.4.1 1620.5.1 9.5.2.5.1 of ASCE 7 9.5.5.7.1 of ASCE 7	C, D, E and F D E and F D, E and F C, D, E and F
2	Reentrant Corners Plan configurations of a structure and its lateral-force-resisting system contain reentrant corners where both projections of the structure beyond a reentrant corner are greater than 15 percent of the plan dimension of the structure in the given direction.	1620.4.1	D, E and F
3	Diaphragm Discontinuity Diaphragms with abrupt discontinuities or variations in stiffness, including those having cutout or open areas greater than 50 percent of the gross enclosed diaphragm area, or changes in effective diaphragm stiffness of more than 50 percent from one story to the next.	1620.4.1	D, E and F
4	Out-of-Plane Offsets Discontinuities in a lateral-force-resistance path, such as out-of-plane offsets of the vertical elements.	1620.4.1 9.5.2.5.1 of ASCE 7 1620.2.9	D, E and F D, E and F B, C, D, E and F
5	Nonparallel Systems The vertical lateral-force-resisting elements are not parallel to or symmetric about the major orthogonal axes of the lateral-force-resisting system.	1620.3.2	C, D, E and F

a. Seismic design category is determined in accordance with Section 1616.

TABLE 1616.5.2—VERTICAL STRUCTURAL IRREGULARITIES

TABLE 1616.5.1.2 VERTICAL STRUCTURAL IRREGULARITIES

	IRREGULARITY TYPE AND DESCRIPTION	REFERENCE SECTION	SEISMIC DESIGN CATEGORY[a] APPLICATION
1a	Stiffness Irregularity—Soft Story A soft story is one in which the lateral stiffness is less than 70 percent of that in the story above or less than 80 percent of the average stiffness of the three stories above.	9.5.2.5.1 of ASCE 7	D, E, and F
1b	Stiffness Irregularity—Extreme Soft Story An extreme soft story is one in which the lateral stiffness is less than 60 percent of that in the story above or less than 70 percent of the average stiffness of the three stories above.	1620.5.1 9.5.2.5.1 of ASCE 7	E and F D, E and F
2	Weight (Mass) Irregularity Mass irregularity shall be considered to exist where the effective mass of any story is more than 150 percent of the effective mass of an adjacent story. A roof that is lighter than the floor below need not be considered.	9.5.2.5.1 of ASCE 7	D, E and F
3	Vertical Geometric Irregularity Vertical geometric irregularity shall be considered to exist where the horizontal dimension of the lateral-force-resisting system in any story is more than 130 percent of that in an adjacent story.	9.5.2.5.1 of ASCE 7	D, E and F
4	In-plane Discontinuity in Vertical Lateral-Force-Resisting Elements An in-plane offset of the lateral-force-resisting elements greater than the length of those elements or a reduction in stiffness of the resisting element in the story below.	1620.4.1 9.5.2.5.1 of ASCE 7 1620.2.9	D, E and F D, E and F B, C, D, E and F
5	Discontinuity in Capacity—Weak Story A weak story is one in which the story lateral strength is less than 80 percent of that in the story above. The story strength is the total strength of seismic-resisting elements sharing the story shear for the direction under consideration.	1620.2.3 9.5.2.5.1 of ASCE 7 1620.5.1	B, C, D, E and F D, E and F E and F

a. Seismic design category is determined in accordance with Section 1616.

TABLE 1609.3.1— EQUIVALENT BASIC WIND SPEED

TABLE 1609.3.1
EQUIVALENT BASIC WIND SPEEDS[a,b,c]

V_{3S}	85	90	100	105	110	120	125	130	140	145	150	160	170
V_{fm}	70	75	80	85	90	100	105	110	120	125	130	140	150

For SI: 1 mile per hour = 0.44 m/s.
a. Linear interpolation is permitted.
b. V_{3S} is the 3-second gust wind speed (mph).
c. V_{fm} is the fastest mile wind speed (mph).

FIGURE 1609—BASIC WIND SPEED (3-SECOND GUST)

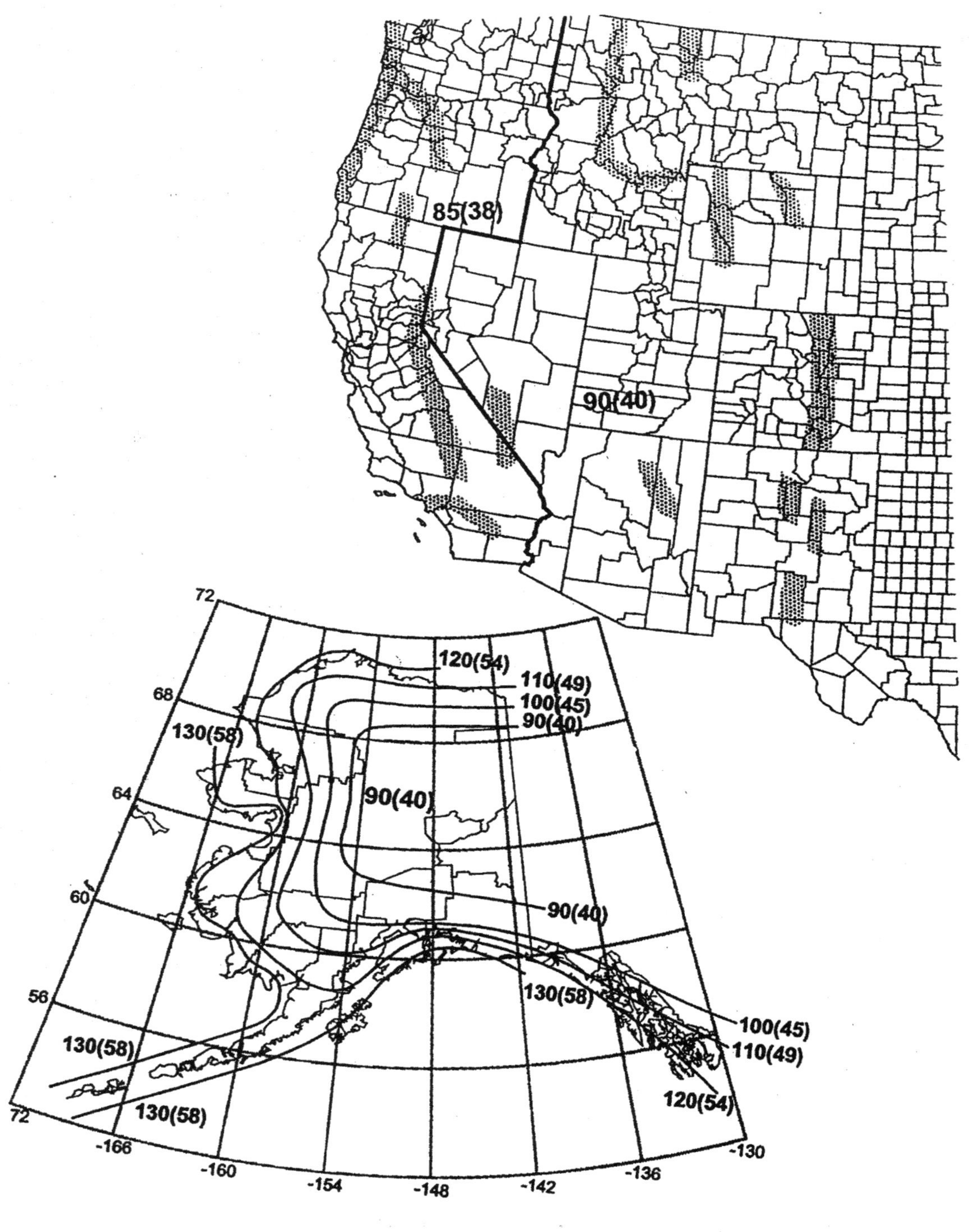

FIGURE 1609
BASIC WIND SPEED (3-SECOND GUST)

FIGURE 1609 (continued)—BASIC WIND SPEED (3-SECOND GUST)

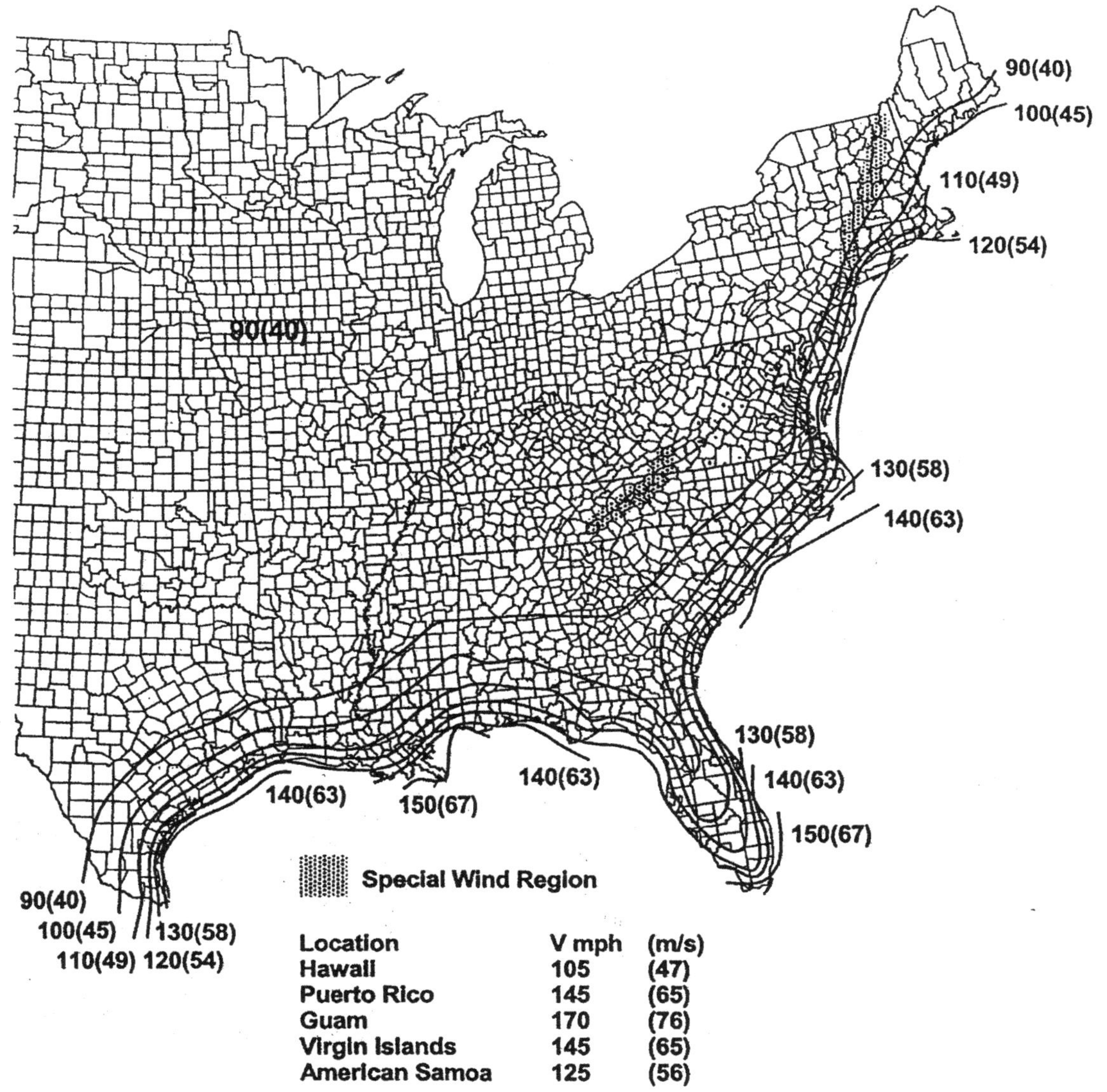

Location	V mph	(m/s)
Hawaii	105	(47)
Puerto Rico	145	(65)
Guam	170	(76)
Virgin Islands	145	(65)
American Samoa	125	(56)

Notes:

1. Values are nominal design 3-second gust wind speeds in miles per hour (m/s) at 33 ft (10 m) above ground for Exposure C category.
2. Linear interpolation between wind contours is permitted.
3. Islands and coastal areas outside the last contour shall use the last wind speed contour of the coastal area.
4. Mountainous terrain, gorges, ocean promontories, and special wind regions shall be examined for unusual wind conditions.

FIGURE 1609—continued
BASIC WIND SPEED (3-SECOND GUST)

GLOSSARY

The following glossary defines a number of structural terms, many of which have appeared on past exams. While this list is by no means complete, it comprises much of the terminology with which candidates should be familiar. You are therefore encouraged to review these definitions as part of your preparation for the structural exams.

A

Acceleration The rate of change of velocity, usually expressed as a fraction or percentage of g, the acceleration of gravity.

Accelerograph A seismological instrument that is normally inoperative, but becomes activated when subject to strong earth motion, records the earth motion, and then shuts off.

Active Pressure The pressure exerted by retained earth against a retaining wall.

Admixture A prepared substance added to concrete to alter or achieve certain characteristics.

Aggregate The chemically inert element of concrete, usually consisting of sand, gravel, and/or other granular material.

Air-Supported Structure A membrane enclosing a pressurized occupied space, which must be held down to its foundation.

Allowable Stress Same as Working Stress.

Arch A curved structure in which the internal stresses are essentially compression.

Axial Load A longitudinal load that acts at the centroid of a member and perpendicular to its cross-section.

B

Balloon Framing A method of framing wood stud walls, in which the studs are continuous for the full height of the building, which is usually two stories.

Base Isolation A method of isolating a structure from the ground by specially designed bearing and dampers that absorb earthquake forces.

Beam A structural member that supports loads perpendicular to its longitudinal axis.

Bearing Wall A wall that supports any vertical load in addition to its own weight.

Bending Moment The algebraic sum of the moments of all forces that are on one side of a given cross-section of a beam.

Braced Frame A vertical truss used to resist lateral forces.

C

C (1) A numerical coefficient used in earthquake design. (2) A standard designation for a structural steel American Standard channel.

Cable Roof A curved structure in which the internal stresses are pure tension.

Camber A curve built into a structural member to compensate for deflection.

Cantilever Beam A beam that is restrained against rotation at one end and free at the other.

Cantilever Footing An exterior column footing joined by a concrete beam to an interior column footing.

Cantilever Wall A retaining wall in which the stem, heel, and toe act as cantilever slabs.

Catenary The shape assumed by a cable when the only load acting on it is its own weight.

Centroid The point in a cross-section where all of the area may be considered concentrated without affecting the moment of the area about any axis.

Chord A perimeter member of a truss.

Coefficient of Thermal Expansion The ratio of unit strain to temperature change, which is constant for a given material.

Collector A member used to collect seismic load from a diaphragm and deliver it to a shear resisting element.

Column A member, usually vertical, that is subject primarily to axial compressive load.

Column Footing A spread footing, generally square or rectangular in plan, used to support a single column.

Combined Footing A footing supporting two or more columns.

Component One of two or more forces that will produce the same effect on a body as a given force.

Composite Beam A steel beam and a concrete slab connected so that they act together as a single structural unit to resist bending stresses.

Composite Deck Steel floor decking with embossed ridges, bonded to a concrete slab so that they act together as a single structural unit.

Compression Stress that tends to shorten a member or crush it.

Compressive Reinforcement Reinforcing steel embedded in the compression face of a reinforced concrete beam.

Concentrated Load A load that acts at one point on a structure.

Concrete A mixture of fine and coarse aggregates, portland cement, and water.

Concurrent Describing the condition when the lines of action of several forces pass through a common point.

Continuous Beam A beam that rests on more than two supports.

Core Test A compression test of hardened concrete which has been cut from the structure.

Counterfort Wall A retaining wall in which the stem and base are connected at intervals by transverse walls called counterforts.

Couple Two forces equal in magnitude, but opposite in direction, and acting at some distance from each other.

Creep Continued deformation of a structural member with time, with no increase of load.

Curing Maintaining concrete at the proper moisture and temperature after it is cast.

Curvature Factor A factor used to modify the allowable unit stress in bending for the curved portion of glued laminated members.

Cylinder Test A test to determine the compressive strength of concrete.

D

Dead Load The vertical load due to the weight of all permanent structural and nonstructural components of a building, such as walls, floors, roofs, and fixed service equipment.

Deflection The movement of a beam from its original location when load is applied to it.

Deformation See Strain.

Diaphragm The horizontal floor or roof system that distributes lateral forces caused by wind or earthquake, by functioning as a horizontal girder.

Dome A roof structure whose shape is that of an arch rotated about its vertical axis to form a curved surface.

Drift The horizontal movement of a structure when subject to wind or earthquake force.

Drilled Caisson An end-bearing pile, the bottom of which may be belled, that is constructed by pouring concrete into a drilled shaft.

Drilled Pile A vertical shaft drilled into the ground and filled with concrete, which supports building loads by skin friction.

Ductility As used in earthquake design, the ability of structural systems and materials to deform and absorb energy, without failure or collapse.

E

E A symbol for modulus of elasticity.

Eccentric Load A longitudinal load that acts at a distance from a member's centroid, thereby producing bending moment in addition to axial stress.

Elastic Limit The unit stress for a material, below which Hooke's Law applies.

End-Bearing Caisson See Drilled Caisson.

Engineering News Formula A dynamic formula used to determine the capacity of driven piles.

Epicenter The projection of the focus on the ground surface.

Equilibrant A force equal in magnitude to the resultant, but opposite in direction and on the same line of action as the resultant.

Equilibrium A state of rest due to balanced forces and balanced moments.

Euler's Equation A basic equation that applies to all columns and gives the maximum stress a slender column can resist without failing by sudden buckling.

Expansive Soil A fine-grained cohesive soil that undergoes large volume changes with changes in moisture content.

F

Factor of Safety The ratio of the ultimate strength of a material to its working stress.

Fault The boundary between adjacent rock plates along which movement may take place during an earthquake.

Fillet Weld A weld placed in the right angle formed by lapping or intersecting plates and generally subject to shear stress.

Fixed End Beam A beam that is restrained (fixed) against rotation at both ends.

Flat Plate A flat slab without column capitals or drop panels.

Flat Slab A concrete slab reinforced in two directions that brings its load directly to supporting columns without any beams or girders, usually requiring column capitals (widened tops of columns) and drop panels (thickened slab around columns).

Flexure Bending.

Focus The location in the earth's crust where rock slippage begins during an earthquake.

Folded Plate A structural roof system consisting of inclined planes that support each other and function as deep beams.

Force A push or pull exerted on an object. The description of a force includes its magnitude, direction, and point of application.

Friction Pile See Drilled Pile.

Frost Line The maximum depth of frost penetration in the ground expected in a given area.

G

Gabled Frame A frame consisting of two columns and two inclined beams that meet at the ridge, in which the joint between each column and beam is rigid.

Girder A main beam that supports secondary beams.

Glued Laminated Beam An assembly of laminations of lumber in which the grain of all the laminations is approximately parallel longitudinally. The laminations are bonded with adhesives and fabricated in accordance with certain accepted standards.

Gravity Wall A retaining wall that depends entirely on its own weight to resist the pressure of the retained earth and provide stability.

Groove Weld A weld placed between two butting pieces of metal to be joined.

Grout A high slump concrete, consisting of portland cement, sand, hydrated water, and sometimes pea gravel.

H

Hooke's Law The physical law that states that up to a certain unit stress, called the elastic limit, unit stress is directly proportional to unit strain.

Hoop A horizontal member that extends around the circumference of a dome.

Hydrostatic Pressure The pressure exerted by a liquid against every surface it contacts.

Hyperbolic Paraboloid A thin shell saddle-shaped surface formed by moving a vertical parabola with downward curvature along and perpendicular to another parabola with upward curvature.

Hypocenter Same as Focus.

I

I (1) A symbol for moment of inertia. (2) The importance factor used in earthquake or wind design.

Impact Hammer Test A nondestructive test of hardened concrete to determine its approximate strength.

J

Joist One of a series of small, closely-spaced beams used to support floor, ceiling, or roof loads.

Joist Girder A shop-fabricated steel truss that supports evenly-spaced steel joists along its top chord.

K

K An effective length factor used in the design of structural steel columns.

Kelly Ball Test A method of measuring the workability of fresh concrete.

Kip A unit of force or weight equal to 1,000 pounds.

Ksi An abbreviation for kips per square inch.

L

L A standard designation for a structural steel angle.

Lamella A roof structure comprising a series of parallel arches, skewed to the axes of the building, which are intersected by another series of skewed arches, so that they interact with each other.

Lateral Load Any horizontal load on a building, including the load from wind or earthquake.

Lift Slab A flat plate cast at grade around columns and then lifted to position with hydraulic jacks.

Line of Action A line parallel to and in line with a force.

Lintel A structural member placed over an opening and supporting construction above.

Liquefaction Transformation of soil into a liquefied state, similar to quicksand, as a result of earthquake vibrations.

Live Load The vertical load caused by the use and occupancy of a building, not including wind, earthquake, or dead loads.

Load A force applied to a body.

M

M A symbol for bending moment.

Mat Foundation A large footing under an entire building, which distributes the building load over the entire area.

Membrane A thin sheet that can resist tension, but cannot resist compression, bending, or shear.

Meridian A curved line on the surface of a dome, usually circular, which is formed by the intersection of a vertical plane with the dome, when the plane passes through the top of the dome.

Method of Joints An analytical method for determining the forces in the members of a truss.

Method of Sections An analytical method for determining the forces in the members of a truss.

Modified Mercalli Scale A scale used to measure the intensity of an earthquake.

Modulus of Elasticity Within the elastic limit, the constant ratio of the unit stress in a material to the corresponding unit strain.

Modulus of Rupture The unit bending stress calculated from the flexure formula, for the maximum bending moment resisted by a beam before rupture.

Moment The tendency of a force to cause rotation about a given point or axis.

Moment Diagram A graphic representation of the value of the bending moment at any point along a beam.

Moment of Inertia The sum of the products obtained by multiplying each unit of area by the square of its distance to the neutral axis.

Moment-Resisting Frame A frame with rigid joints, in which the members and joints are capable of resisting vertical and horizontal forces primarily by flexure.

N

Natural Period of Vibration The time it takes for a structure to go through one complete back-and-forth motion under the action of dynamic loads.

Negative Moment Bending moment that produces tension in the upper part of a beam and compression in the lower part.

Neutral Axis The line on a beam cross-section that has zero bending stress when the beam is loaded.

Nonbearing Wall A wall that supports no vertical load other than its own weight.

O

Open Web Steel Joist A shop-fabricated, lightweight steel truss used to span between main members or bearing walls and support roof or floor loads.

Overhanging Beam A beam that rests on two or more supports and has one or both ends projecting beyond the support.

P

P-Delta Effect The secondary effect on frame members produced by vertical loads acting on a building frame that is laterally displaced by earthquake loads.

Pile An underground wood, concrete, or steel member, usually vertical, and usually driven into place, that is used to support building loads.

Plate Girder An assembly of steel plates, or plates and angles, which are fastened together to form an integral member.

Plate Tectonics The theory that explains earthquake phenomena.

Platform Framing A method of framing wood stud walls in which the studs are one story in height and the floor joists bear on the top plates of the wall below.

Point of Inflection The point in a beam or other flexural member where the bending moment changes sign and has a value of zero.

Portland Cement The finely-ground material used as the binder for structural concrete.

Positive Moment Bending moment that produces compression in the upper part of a beam and tension in the lower part.

Posttensioning A method of prestressing concrete in which the concrete is cast and then the steel tendons stressed by jacking.

Prestressed Concrete Concrete that is permanently loaded so as to cause stresses opposite in direction from those caused by dead and live loads.

Pretensioning A method of prestressing concrete in which the tensile force is put into high strength steel wires before the concrete is cast.

Proctor Test A laboratory compaction test to determine the optimum moisture content and density for a soil.

Psf An abbreviation for pounds per square foot.

Psi An abbreviation for pounds per square inch.

Punching Shear Two-way shear that occurs in a flat slab, spread footing, or pile cap.

Purlin A regularly spaced roof beam that spans between girders or trusses.

R

r A symbol for radius of gyration.

R_w A numerical coefficient used in earthquake design.

Radiographic Inspection Nondestructive testing of welded joints using x-rays and gamma rays.

Radius of Gyration A term used in column design equal to $\sqrt{I/A}$, where I is the moment of inertia of a member, and A is its cross-sectional area.

Reactions Forces acting at the supports of a structure which hold the structure in equilibrium.

Redundancy The ability of a part of a structural system to redistribute loads to other parts of the system.

Redundant Member Any truss member not necessary for stability.

Reinforced Brick Masonry A type of wall construction consisting of brick units, usually two tiers, with a solidly grouted space between each in which vertical and horizontal reinforcing bars are placed.

Reinforced Concrete Block Masonry A type of wall construction consisting of hollow concrete masonry units, with certain cells continuously filled with grout in which reinforcing bars are embedded.

Resolving Forces Replacing a force with two or more other forces (components) that will produce the same effect on a body as the original force.

Response Spectrum A curve that shows the maximum acceleration of a series of idealized structures when subject to an earthquake.

Resultant One force that will produce the same effect as two or more other forces.

Retaining Wall A wall that resists the lateral pressure of retained earth or other material.

Richter Scale A logarithmic scale used to measure earthquake magnitudes.

Rigid Frame Same as Moment-Resisting Frame.

Rigidity Same as Stiffness.

S

S (1) A symbol for section modulus. (2) A standard designation for a structural steel I beam, also known as an American Standard beam. (3) The site coefficient for soil characteristics used in earthquake design.

Section Modulus The ratio of the moment of inertia of a beam (I) to the distance from its neutral axis to the most remote fiber (c). Thus, section modulus (S) = I/c.

Shear Stress that tends to make two members, or two parts of a member, slide past each other.

Shear Diagram A graphic representation of the value of the vertical shear at any point along a beam.

Shear Wall A wall designed to resist lateral forces parallel to itself caused by wind or earthquake.

Simple Beam A beam that rests on a support at each end.

Size Factor A factor used to reduce the allowable bending stress for wood beams deeper than 12 inches.

Slenderness Ratio The ratio l/r used in column design, where l is the length and r is the radius of gyration.

Slump Test A test for mixed concrete to determine consistency and workability.

Space Frame A series of trusses that intersect in a consistent grid pattern and are rigidly connected at their points of intersection.

Spiral Column A reinforced concrete column, usually square or round, containing longitudinal reinforcing bars enclosed by a closely-spaced continuous steel spiral.

Statical Moment The product of an area and the distance from the centroid of the area to a given axis.

Statically Determinate Describing a structure whose reactions can be determined from the equations of equilibrium.

Statically Indeterminate Describing a structure whose reactions cannot be found from the equations of equilibrium only, but requires additional equations.

Stiffness Resistance to deformation.

Stirrup A vertical steel bar, usually U-shaped, used to reinforce a reinforced concrete beam where the shear stresses are excessive.

Strain The deformation (change in size) of a body caused by external forces.

Strap Footing Same as Cantilever Footing.

Strength Design The method generally used for reinforced concrete design, formerly called ultimate strength design.

Stress An internal force in a body that resists an external force.

Stress Diagram A graphical method for determining the force in the members of a truss.

Stressed Skin A structural system consisting of spaced members solidly sheathed on one or both sides, in which the sheathing forms the flanges and resists flexure while the spaced members comprise the webs and resist shear.

Stud Wall A wall consisting of small closely-spaced members (studs) usually sheathed on both faces with a wall material.

Surcharge Increased earth pressure against a retaining wall caused by vertical load behind the wall or a sloping ground surface.

T

T-Beam A reinforced concrete beam consisting of a portion of the slab and the integrally constructed beam, which act together.

Tapered Girder A plate girder having a tapered profile, usually varying from minimum depth at the supports to maximum depth at mid-span.

Tension Stress that tends to stretch a member or pull it apart.

Thin Shell A structure with a curved surface that supports load by tension, compression, and shear in the plane of its surface, but which is too thin to resist bending stresses.

Tied Column A reinforced concrete column, usually square or rectangular, containing longitudinal reinforcing bars and separate lateral ties.

Tilt-Up Wall A reinforced concrete wall that is precast at the job site, usually in a flat position, and later tilted up and set into place.

Torsion The rotation caused in a diaphragm by lateral load from wind or earthquake, when the center of mass does not coincide with the center of rigidity.

Tributary Area The floor or roof area supported by an individual structural member.

Truss A jointed structure designed to support vertical or horizontal loads and composed generally of straight members forming a number of triangles.

Trussed Rafter A prefabricated lightweight wood truss used to support roof loads for dwellings and other small structures.

Tsunami An ocean wave produced by displacements of the ocean bottom as a result of earthquake or volcanic activity.

Tubular System A structural system used in tall buildings, consisting of closely-spaced columns at the perimeter connected by deep spandrel beams, which acts like a tube that cantilevers from the ground when subject to lateral wind or earthquake loads.

Ultimate Strength The maximum unit stress that can be developed in a material.

Ultrasonic Testing Nondestructive testing of welded joints using high-frequency sound waves.

Uniformly Distributed Load A beam loading of constant magnitude per unit of length.

V

V (1) A symbol for vertical shear. (2) The total lateral earthquake force at the base of a structure.

Vault A series of arches placed side-by-side to form a continuous structure.

Vertical Shear The algebraic sum of the forces that are on one side of a given cross-section of a beam.

Vibratory Compactor A machine used primarily for the compaction of granular soils, such as sands.

Vierendeel Truss A truss with no diagonals.

Volume Factor A factor used to reduce the allowable bending stress for glued laminated beams, based on width, depth, and span.

W

W (1) The total dead load used in earthquake design. (2) A standard designation for a structural steel wide flange shape.

Wall Footing A continuous spread footing supporting a uniformly loaded wall.

Water-Cement Ratio The ratio of water to cement in a concrete mix, the main factor that determines concrete strength.

Web Members The interior members of a truss, which connect the chords.

Wind Bent A frame used to resist lateral forces from wind.

Workability The ease with which concrete can be placed and consolidated in the forms.

Working Stress The maximum unit stress permissible in a structural member.

Working Stress Design The theory used for most reinforced concrete design until the middle 1960s.

Y

Yield Point The unit stress at which a material deforms with no increase in load.

Z

Z A numerical coefficient used in earthquake design which depends on the seismic zone in which the building is located.

BIBLIOGRAPHY

The Architect Registration Examination (ARE) is closed book; no reference material is permitted for any part of the exam. If you need any reference material for the structural test, you will be able to access it during the test.

We suggest that prior to taking the test, you obtain some of the reference books that may be excerpted, in order to become more familiar with them and thereby save valuable time at the exam.

The reference books that have often had sections reproduced for reference during the test include the *AISC Manual*, the *Uniform Building Code*, and the *Standard Specifications for Steel Joists*.

These reference books are listed below, along with other useful books for those candidates wishing to do additional studying or review.

ACI Code (Building Code Requirements for Reinforced Concrete ACI 318)
American Concrete Institute
P.O. Box 9094
Farmington Hills, MI 48333

Architects and Earthquakes
AIA Research Corporation
1735 New York Ave., NW
Washington, DC 20006

Manual of Steel Construction,
Ninth Edition
American Institute of Steel Construction
One East Wacker Drive
Chicago, IL 60601

National Design Specification for Wood Construction
American Forest & Paper Association
1111 19th Street, NW
Washington, DC 20036

Parker Handbooks
Wiley

1. Simplified Design of Structural Steel
2. Simplified Mechanics and Strength of Materials
3. Simplified Design of Reinforced Concrete
4. Simplified Engineering for Architects and Builders

Standard Specifications for Steel Joists
Steel Joist Institute
1205 48th Ave. North
Myrtle Beach, SC 29577

Structure in Architecture
Salvadori and Heller
Prentice-Hall

Timber Construction Manual
American Institute of Timber Construction
Englewood, CO
Wiley

Uniform Building Code
International Conference of Building Officials
5360 South Workman Mill Road
Whittier, CA 90601

QUIZ ANSWERS

Lesson 1

1. **D** If you refer to UBC Table 23-I-J-l on page 24, you will note that the allowable shear in plywood diaphragms varies with plywood thickness, width of framing members, and nail spacing. The direction of framing members does not affect the strength of a blocked diaphragm, that is, one having blocking at the plywood panel edges.

2. **B** Moment-resisting frames are the most ductile lateral load resisting system, and are therefore most effective in resisting earthquakes, particularly for tall buildings. Locating these frames at the building's perimeter tends to minimize the potentially damaging effects of torsion.

3. **D** A, B, and C are correct (see pages 12 and 20). D is incorrect; torsion is the rotation in a diaphragm that occurs when the center of mass does not coincide with the center of rigidity.

4. **D** II and III are correct and I is incorrect (see page 3). IV is also incorrect, since live loads are vertical gravity loads unrelated to earthquake loads.

5. **B** The system described is system 2.3.a, which, according to Table 16-N on page 16, has an R_w value of 8. The seismic zone is irrelevant.

6. **B** See pages 27 and 28.

7. **C** One of the factors in the formula for the period of a structure (T) is the height of the building (see page 9). The taller the building, the longer the period.

8. **D** See page 11.

9. **B** The intent of the Uniform Building Code earthquake provisions is to make buildings capable of resisting major earthquakes without collapse, although some structural and nonstructural damage may occur (see page 6).

10. **A** See page 3.

Lesson 2

1. **A** The wind pressure on a building acts inward on the windward wall (I), outward on the leeward wall (II), and upward on a flat roof (III). See page 54 and Table 16-H on page 59.

2. **B** Drift and overturning moment (I and IV) are considered in wind design, while liquefaction and tsunamis (II and III) are phenomena associated with earthquakes.

3. **B** Wind pressure $p = C_e C_q q_s I$.
For exposure C, $C_e = 1.06$ (see Table 16-G on page 52).
C_q for the windward wall using Method 1 = 0.8 (see Table 16-H on page 53). q_s for a 70 mph wind speed = 12.6 psf (see Table 16-F on page 54).
$p = 1.06(0.8)(12.6)(1.0) = 10.68$ psf

4. **C** The UBC wind load coefficients account for all of the factors shown, except tornadoes. (I, II, and IV are correct, III is incorrect.)

5. **B** For wind load, the north and south walls span vertically between the ground and the roof diaphragm. The total wind load delivered to the roof diaphragm is therefore 20 psf × 12 ft./2 × 100 ft. = 12,000#. Half of this load goes into the east wall and half to the west wall, or 6,000# to each wall.

6. **A** The maximum diaphragm shear is equal to the wind load transferred to each end wall divided by the width of the diaphragm, or 6,000 #/50 ft. = 120#/ft.

7. **C** See page 58.

8. **C** The factor of safety against overturning is equal to the dead load resisting moment

divided by the overturning moment = 500 ft.-kips ÷ 400 ft.-kips = 1.25. This is inadequate, since the UBC requires a factor of safety of at least 1.5. Although the factor of safety could be increased by making the wall longer and/or heavier (B), this is not generally feasible. A better solution is to provide sufficient anchorage from the wall to the foundation (C), so that the dead load of the foundation can be used to help resist the overturning moment.

9. **D** The tubular concept is discussed in Lesson 1, since tubular systems are used to resist lateral loads from either earthquake or wind. For very tall buildings, this system is often economical (see page 34).

10. **A** The load combinations shown in B, C, and D must be considered in the design of building components. The combination shown in A is not considered in the design, since the likelihood that the roof live load will be acting on the structure during a windstorm is very remote.

EXAMINATION DIRECTIONS

The examination on the following pages should be taken when you have completed your study of all the lessons in this course. It is designed to simulate the Lateral Forces divisions of the Architect Registration Examination. Many questions are intentionally difficult in order to reflect the pattern of questions you may expect to encounter on the actual examination.

You will also notice that the subject matter for several questions has not been covered in the course material. This situation is inevitable and, thus, should provide you with practice in making an educated guess. A few questions may appear ambiguous, trivial, or simply unfair. This too, unfortunately, reflects the actual experience of the exam and should prepare you for the worst you may encounter.

Answers and complete explanations will be found on the pages following the examination, to permit self-grading. **Do not look at these answers until you have completed the entire exam.** Once the examination is completed and graded, your weaknesses will be revealed, and you are urged to do further study in those areas.

Please observe the following directions:

1. If a question requires you to refer to a table or chart in the course, it will so state. Otherwise, please do not use any reference material.
2. Allow about 1-1/2 hours to answer all questions. Time is definitely a factor to be seriously considered.
3. Read all questions *carefully* and mark the appropriate answer on the answer sheet provided.
4. Answer all questions, even if you must guess. Do not leave any question unanswered.
5. If time allows, review your answers, but do not arbitrarily change any answer.
6. Turn to the answers only after you have completed the entire examination.

GOOD LUCK!

EXAMINATION ANSWER SHEET

Directions: Read each question and its lettered answers. When you have decided which answer is correct, blacken the corresponding space on this sheet. After completing the exam, you may grade yourself; complete answers and explanations will be found on the pages following the examination.

1. Ⓐ Ⓑ Ⓒ Ⓓ
2. Ⓐ Ⓑ Ⓒ Ⓓ
3. Ⓐ Ⓑ Ⓒ Ⓓ
4. Ⓐ Ⓑ Ⓒ Ⓓ
5. Ⓐ Ⓑ Ⓒ Ⓓ
6. Ⓐ Ⓑ Ⓒ Ⓓ
7. Ⓐ Ⓑ Ⓒ Ⓓ
8. Ⓐ Ⓑ Ⓒ Ⓓ
9. Ⓐ Ⓑ Ⓒ Ⓓ
10. Ⓐ Ⓑ Ⓒ Ⓓ
11. Ⓐ Ⓑ Ⓒ Ⓓ
12. Ⓐ Ⓑ Ⓒ Ⓓ
13. Ⓐ Ⓑ Ⓒ Ⓓ
14. Ⓐ Ⓑ Ⓒ Ⓓ
15. Ⓐ Ⓑ Ⓒ Ⓓ
16. Ⓐ Ⓑ Ⓒ Ⓓ
17. Ⓐ Ⓑ Ⓒ Ⓓ
18. Ⓐ Ⓑ Ⓒ Ⓓ
19. Ⓐ Ⓑ Ⓒ Ⓓ
20. Ⓐ Ⓑ Ⓒ Ⓓ
21. Ⓐ Ⓑ Ⓒ Ⓓ
22. Ⓐ Ⓑ Ⓒ Ⓓ

FINAL EXAMINATION

1. Building A has a fundamental period of vibration of one second. Building B has a fundamental period of vibration of two seconds. All other factors for the two buildings are equal. Select the correct statement.

   A. Building A has a greater seismic force.

   B. Building B has a greater seismic force.

   C. Both buildings have equal seismic forces.

   D. The information given is insufficient to determine which building has the greater seismic force.

2. 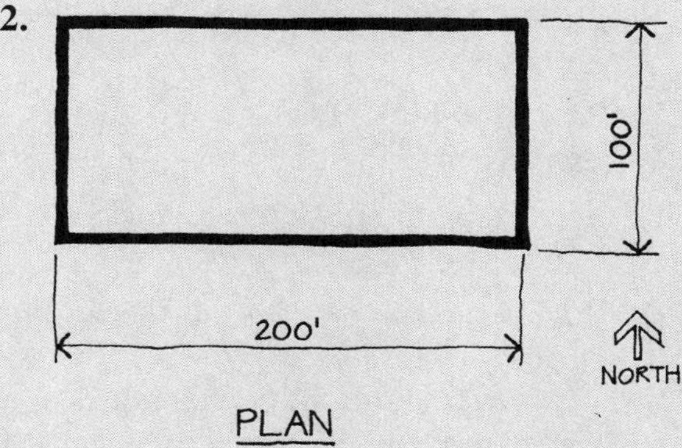

   The building shown above has rigid horizontal diaphragms that transfer seismic forces to vertical shear walls at the building's perimeter. The center of mass coincides with the center of rigidity. In the seismic design of the building, horizontal torsional moments

   A. are ignored.

   B. must be provided for by assuming the mass displaced 10 feet for north-south loads and 5 feet for east-west loads.

   C. must be provided for by assuming the mass displaced 5 feet in each direction.

   D. must be provided for by assuming the mass displaced 10 feet in each direction.

3. Select the correct statements about damage from earthquake forces.

   I. Tall buildings are particularly susceptible to damage from slow rocking motions of an earthquake.

   II. Damage to elevators in tall buildings is often significant.

   III. One-story wood frame dwellings perform comparatively well, and substantially better than two-story dwellings.

   IV. Heavy damage from earthquakes may be expected where soft alluvial soils meet firmer soils.

   V. Major damage and hazard to life is generally from ground shaking rather than surface faulting.

   A. I, II, and IV

   B. III and V

   C. I, II, III, and IV

   D. I, II, III, IV, and V

4. Which of the following buildings best expresses the structural concept of the "bundled tube?"

   A. Sears Tower

   B. John Hancock Building

   C. CBS Building

   D. Knights of Columbus Building

5. In the building shown below in plan, earthquake forces in the north-south direction and torsion are resisted by bents 1 and 2. Bents 2 are more rigid than bents 1. Select the correct statement.

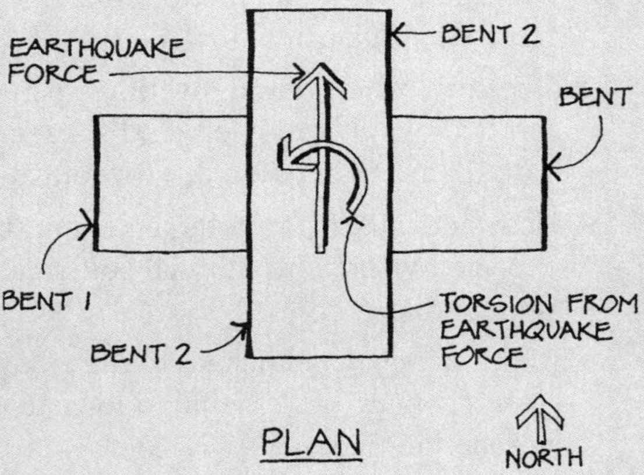

A. Bents 1 will resist most of the direct earthquake shear and bents 2 will resist most of the torsional earthquake shear.

B. Bents 1 will resist all of the direct and torsional earthquake shear.

C. Bents 2 will resist all of the direct and torsional earthquake shear.

D. Bents 1 will resist most of the torsional earthquake shear and bents 2 will resist most of the direct earthquake shear.

6. The braced frame below resists a wind force of 6 kips as shown. Assuming that the braces can resist tension only, what are the forces in braces 1 and 2?

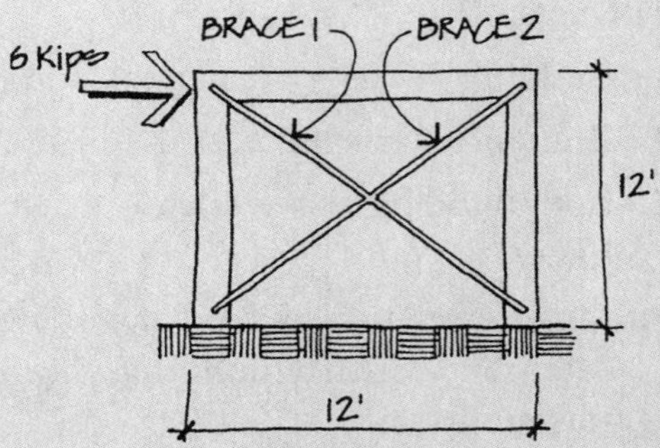

A. Brace 1 = 6.0 kips tension, brace 2 = zero

B. Brace 1 = zero, brace 2 = 6.0 kips tension

C. Brace 1 = 8.5 kips tension, brace 2 = zero

D. Brace 1 = zero, brace 2 = 8.5 kips tension

7. In the braced frame of the previous question, what is the force in the right column caused by the wind?

A. 6 kips compression

B. 6 kips tension

C. 8.5 kips compression

D. 8.5 kips tension

8. Select the INCORRECT statement.

   A. Nonstructural walls can drastically affect the performance of a building in an earthquake.

   B. A building with many partitions in the upper stories and a relatively open first story may be expected to perform well in an earthquake.

   C. Reinforced concrete columns subject to earthquake forces often fail because of inadequate ties.

   D. An irregular floor plan tends to increase torsional stresses from earthquake forces.

9. Select the INCORRECT statement about rigid frames.

   A. A rigid frame can resist both gravity and wind loads.

   B. Steel or reinforced concrete may be used in rigid frame construction.

   C. A rigid frame is more rigid than a braced frame of comparable dimensions.

   D. The joints of a rigid frame are capable of resisting bending moment.

10.

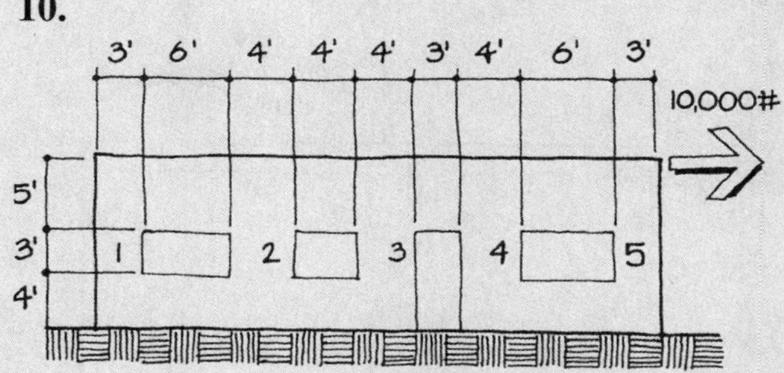

In the wall shown above,

   A. piers 2, 3, and 4 resist 2,300# each and piers 1 and 5 resist 1,500# each.

   B. piers 1 and 5 resist 2,300# each and piers 2, 3, and 4 resist 1,800# each.

   C. piers 1 and 5 resist 1,500# each, pier 2 resists 3,000#, and piers 3 and 4 resist 2,000# each.

   D. each pier resists 2,000#

11.

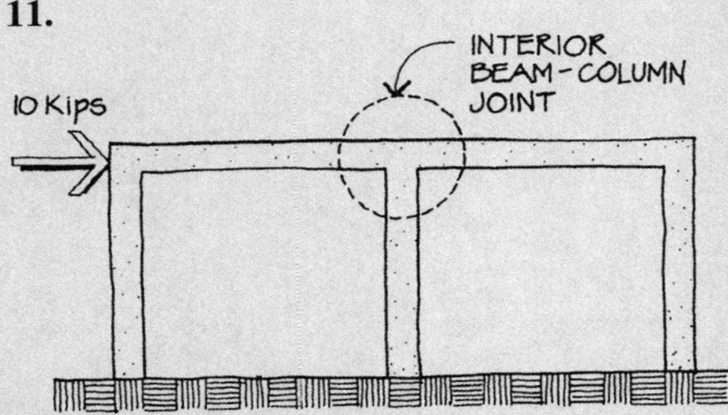

A wind load of 10 kips is resisted by a moment-resisting frame as shown above. Which of the choices below correctly shows the internal forces in the members at the interior beam-column joint? (T = tension, C = compression.)

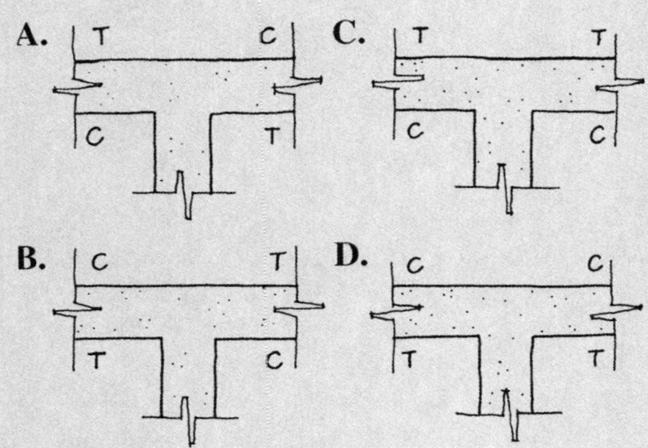

12. In designing a building for earthquake forces in accordance with the Uniform Building Code, which of the following are NOT considered?

I. The building's orientation

II. The subsoil conditions

III. The geographical location of the building

IV. The dead weight of the building

V. The type of lateral load resisting system

A. I and II **C.** II, III, IV, and V

B. I, II, and III **D.** I only

13. For the water tank supported on the four-legged braced frame shown below, what is the maximum vertical force at the bottom of column A due to gravity and seismic loads? V = 0.3W.

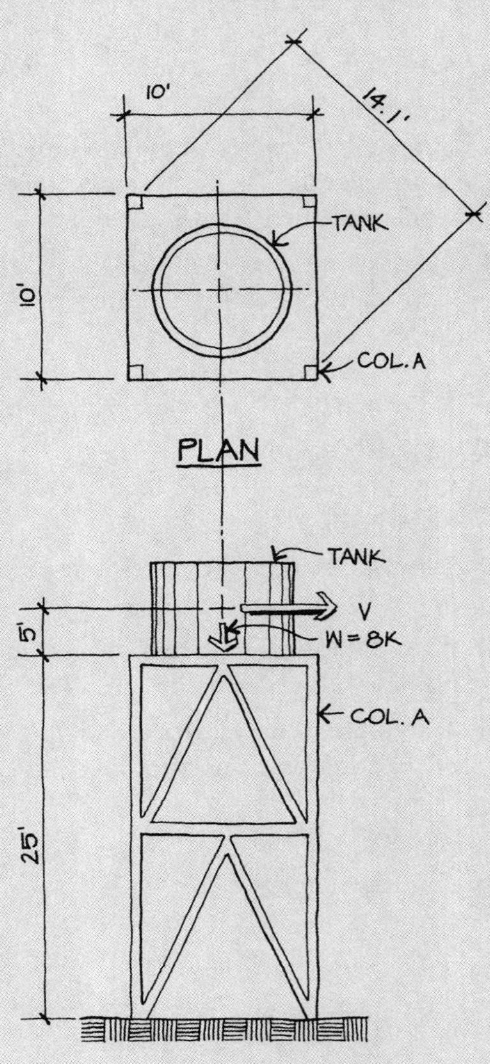

A. 11.2 kips **C.** 5.6 kips

B. 7.1 kips **D.** 5.1 kips

14. If the wind speed doubles, the wind pressure

A. doubles.

B. increases fourfold.

C. remains the same.

D. decreases by one half.

15. Which of the following statements about live load is correct?

A. Live load is not combined with wind or earthquake loads in the structural design of buildings.

B. Live load is the load superimposed by the use and occupancy of the building, not including the wind load, earthquake load, or dead load.

C. The allowable stresses may be increased 25 percent for structural members supporting dead load plus live load.

D. Roof live load is assumed to act concurrently with wind, snow, or earthquake loads.

16. A seven-story office building in Seismic Zone 4 has special moment-resisting frames. What is the seismic base shear as a percentage of the building weight? Since the period of the building and the soil conditions are not known, use the maximum value of C = 2.75. You may use the UBC tables in Lesson 1.

A. 6.9%

B. 9.2%

C. 11.0%

D. 18.3%

17. Floor plans of four reinforced concrete buildings are shown below. Which would experience the greatest amount of torsion when resisting lateral forces?

A.

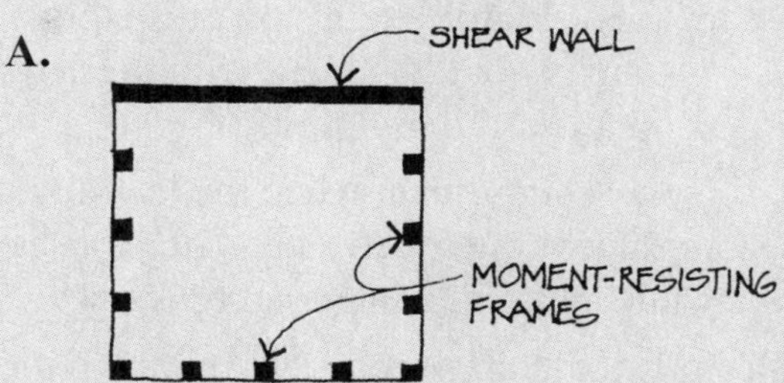

B.

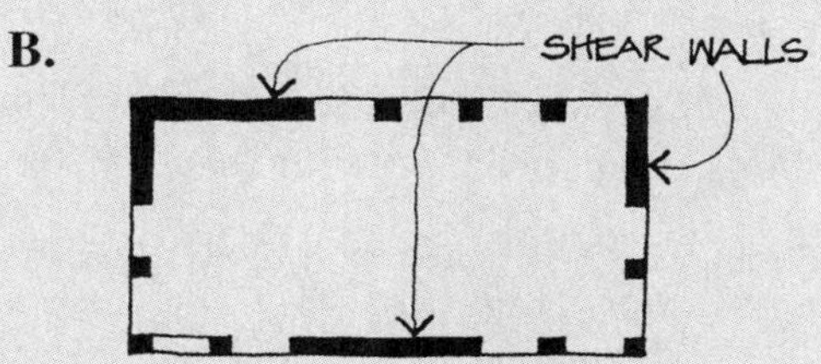

C.

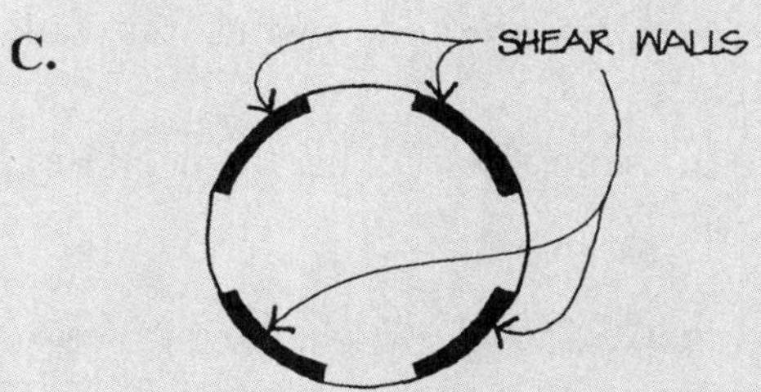

D.

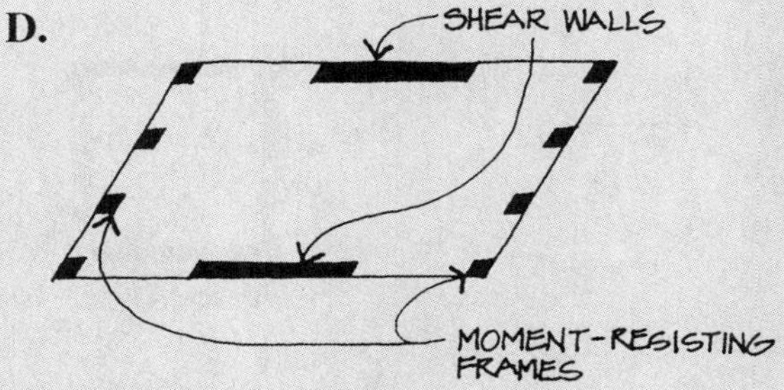

18. In earthquake design of buildings, torsion

A. occurs only in buildings which are circular in plan.

B. occurs when the center of resistance of the primary lateral load resisting system is not coincident with the center of mass of the building.

C. is most significant when the lateral load resisting system is a dual system of shear walls and moment-resisting frames.

D. must be considered in buildings with flexible diaphragms.

19. The wind pressures used in the design of buildings in accordance with the UBC

A. do not account for the effects of hurricanes.

B. increase when there are buildings or surface irregularities close to the site.

C. act both inward and outward on wall surfaces.

D. are the same for wall elements and wall corners.

20. In earthquake design, which of the following are examples of plan irregularities?

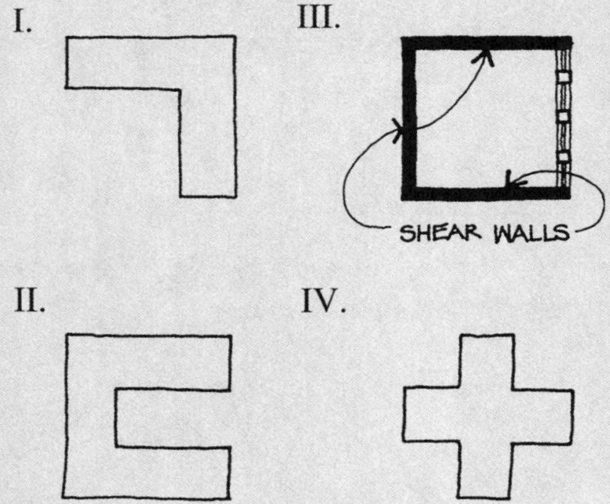

A. III only

B. I and III

C. I and II

D. I, II, III, and IV

21. Your client has proposed that an existing large warehouse, designed and built under the requirements of the Building Code, be converted to a meeting hall for 600 persons. To meet requirements, you must show that the building's lateral force resisting system is adequate to resist how much more lateral force than its original design forces?

A. No additional lateral force

B. 15 percent more wind force only

C. 15 percent more wind force, 25 percent more seismic force

D. 25 percent more seismic force only

22. In the detail shown below, what is the purpose of anchor A?

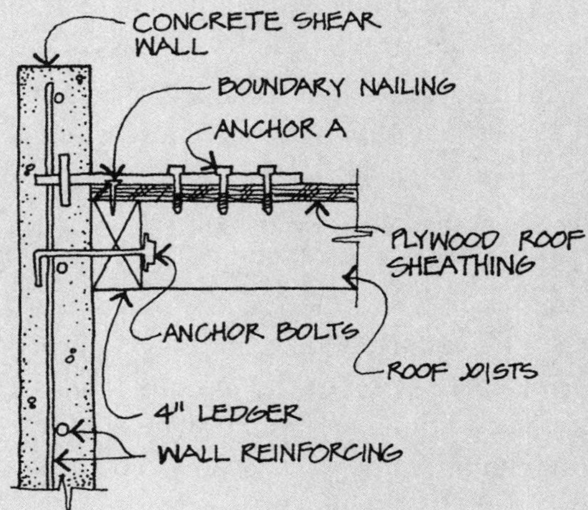

A. Transfers vertical load from roof joists to wall.

B. Transfers lateral load from roof diaphragm to wall.

C. Anchors wall to roof diaphragm for seismic or wind forces perpendicular to wall.

D. Connects plywood panels together.

The examination answers and explanations will be found on the following pages.

Do not look at the answers until you have completed the exam.

EXAMINATION ANSWERS

1. **A** In general, the longer the fundamental period, the lower the acceleration and the resulting seismic force. Conversely, the shorter the period, the greater the seismic force. The correct answer is therefore A.
2. **B** Refer to page 100 of this course or UBC Section 1628.5. To account for various uncertainties, accidental torsion must be provided for by assuming the mass displaced from the calculated center of mass in each direction a distance equal to 5 percent of the building dimension perpendicular to the direction of the force. Thus, for north-south loads, $0.05 \times 200 = 10$ feet, and for east-west loads, $0.05 \times 100 = 5$ feet.
3. **D** All of the statements are true. Tall buildings have a long fundamental period of vibration, and therefore tend to be in resonance with the long period of a slow rocking motion (I). Elevator cables and the containment of elevator counterweights in tall buildings were typically damaged in the San Fernando earthquake of 1971 (II). The typical one-story wood frame dwelling is particularly suited to survive strong shaking without serious danger to occupants, although such houses may be damaged. Two-story houses typically do not perform as well as one-story houses (III). In the San Fernando earthquake, intensified damage occurred at the discontinuity between soft alluvial soils and the much firmer soils of the foothills (IV). Most of the deaths in recent earthquakes were due to ground shaking, rather than surface faulting (V). Building directly on known active faults should be avoided, but it is more important to design structures to resist ground shaking with minimum hazard to occupants and without excessive damage.
4. **A** In the "tube" concept, the entire structure acts like an immense, hollow tubular column which cantilevers out of its foundation under the action of wind loads. The Sears Tower (correct answer A) consists of a "bundle" of nine tubes that terminate at varying heights, thus expressing the "bundled tube" concept visually.
5. **D** Bents 1, being furthest from the center of the building, will resist most of the shear from torsion. Bents 2, being stiffer, will resist most of the direct shear. The correct answer is therefore D.
6. **D** A braced frame is essentially a vertical truss which resists horizontal loads. We cut a section through the braces, as shown in the right column.

   $\Sigma H = 0$

   $6 - F_{1H} - F_{2H} = 0$

   The internal force in brace 1 acts toward the upper left joint, which means that the brace is in compression. But the braces can only resist tension, and therefore brace 1 is ineffective and has zero force.

   Since $6 - F_{1H} - F_{2H} = 0$

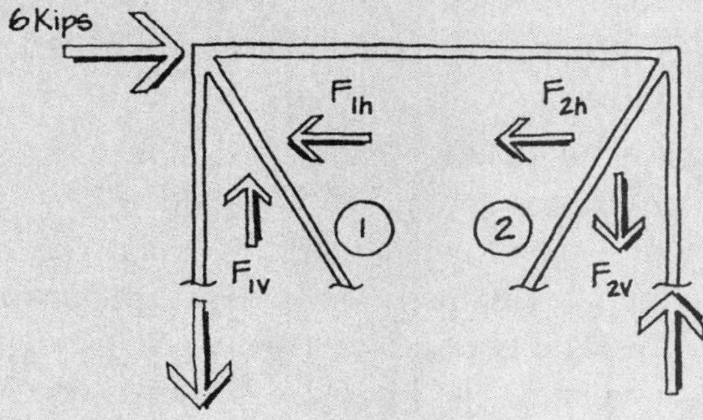

   $6 - 0 - F_{2H} = 0$

   $F_{2H} = 6$

   Since the brace is at 45° to the horizontal, its force is $F_{2H} \div \cos 45° = 6 \div 0.707 = 8.5$ kips.
7. **A** Take moments about the bottom of the left column. 6 kips (12 ft.) – force in right column (12 ft.) = 0. Force in right column = 6 kips. Since the foundation acts upward on the column, the column is in compression.

8. **B** Statement A is true. Although nonstructural walls are assumed to provide no resistance to a building's earthquake forces, they can in fact affect the way the building performs in an earthquake. Statement B is false. The sudden change from the stiff upper stories to the flexible first story creates problems in the building's performance under earthquake loading. Statements C and D are both true. The only incorrect statement is B, and it is therefore the correct answer.

9. **C** A rigid frame consists of beams and columns rigidly connected together so that the joints are capable of resisting bending moment (D is correct). Rigid frames can efficiently resist both vertical and horizontal forces (correct statement A). Their greatest use is in multistory construction, and they may be constructed of either structural steel or reinforced concrete (B is correct). Statement C is the answer we are seeking. Despite its name, a rigid frame is generally less rigid than a comparable braced frame. In other words, a rigid frame tends to deflect more than a braced frame.

10. **A** The 10,000# lateral load will be distributed among the five piers in proportion to their rigidities, where the rigidity is defined as resistance to deflection. The rigidity of a pier depends on its height and depth. The lower the height-to-depth ratio, the greater the rigidity. Piers 2, 3, and 4 each have the same rigidity, based on a height of three feet and a depth of four feet, and therefore a height-to-depth ratio of 3/4 = 0.75. Piers 1 and 5 have equal rigidities based on a height of three feet and a depth of three feet, and a height-to-depth ratio of 3/3 = 1.0. The rigidity of piers 2, 3, and 4 is therefore greater than that of piers 1 and 5. Thus, the lateral loads resisted by piers 2, 3, and 4 are equal, and the lateral loads resisted by piers 1 and 5 are equal, but smaller than those resisted by piers 2, 3, and 4. Only answer A meets these criteria.

11. **A** We isolate the joint in question, as shown on the following page. Since the wind load acts to the right, the internal force in the column must act to the left. That causes a clockwise moment in the column, which must be balanced by counterclock-wise moments in the beams. The shear in the left beam therefore acts down, while that in the right beam acts up. The face of the beam or column that tends to open up because of the

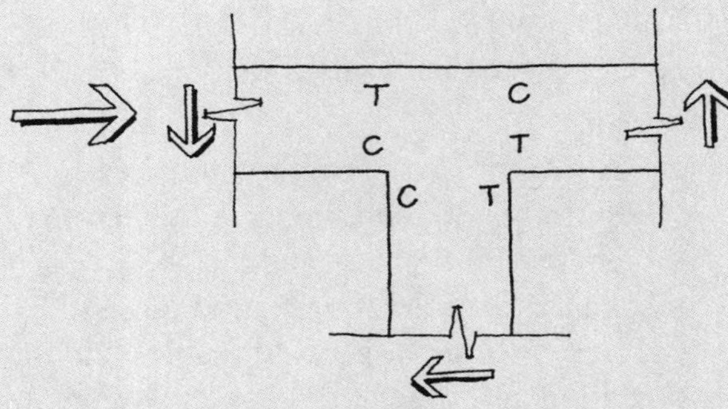

moments caused by these forces is in tension (T), while the face tending to close or crush is in compression (C), as shown. A is therefore the correct answer.

12. **D** The total lateral earthquake force on a structure is equal to $\frac{ZIC}{R_w}$ W. In this formula, Z is the seismic zone factor, which depends on the building's location (III), I is the importance factor, C is a coefficient that is based partially on the soil characteristics at the site (II), R_W is a coefficient that depends on the building's lateral load resisting system (V), and W is the total dead load (IV). The building's orientation (I) is not considered in earthquake design, making D the correct answer.

13. **B** Questions involving overturning moment from wind or earthquake forces have appeared on recent exams, and candidates are therefore advised to become

familiar with this concept. For the water tank in this question, the gravity load on each of the four legs is one-quarter of the total load or 8 kips/4 = 2 kips. The seismic force on the structure is given as V = 0.3 W = 0.3 × 8 kips = 2.4 kips. The overturning moment on the structure is then 2.4 kips × (25 ft. + 5 ft.) = 72 ft.-kips. If the seismic force acts parallel to the main axes of the structure, all four legs resist the overturning moment and the vertical force due to overturning on each leg is 72 ft.-kips/10 ft. × 2 legs = 2.6 kips (up or down). However, the seismic force may act in any direction, and if it acts along a 45° diagonal, then only two legs resist the overturning moment. In that case, the vertical force on two diagonal legs, due to the overturning moment, is 72 ft.-kips/14.1 ft. = 5.1 kips (up or down). The maximum vertical force due to both gravity and seismic overturning is 2 kips + 5.1 kips = 7.1 kips (correct answer B).

14. B The wind pressure varies as the square of the velocity, and therefore doubling the wind velocity increase the wind pressure fourfold.

15. B The correct definition of live load is given in answer B. Floor live load is assumed to act concurrently with wind or earthquake loads (A is incorrect), but roof live load is not (D is incorrect). C is also incorrect; there is usually no basis for increasing the allowable stresses for a member supporting dead plus live load.

16. B The seismic base shear is equal to $\frac{ZIC}{R_w}$ W. Z in Seismic Zone 4 is 0.40 (see Table 16-I on page 8). I is 1.0, since this office building is not an essential or hazardous building. C is given as 2.75. The building has special moment-resisting frames (system 3.1.a or 3.1.b in Table 16-N on page 16), and such systems have an R_W value of 12.

$$\frac{ZIC}{R_w}W = \frac{0.40 \times 1.0 \times 2.75}{12}W$$

$$= 0.092W, \text{ or } 9.2\% \text{ of } W.$$

17. A Torsional forces on a building result when lateral forces act on a building that is not symmetrical with respect to rigidity. In the four plans in this question, the lateral forces from wind or earthquake act through the center of gravity of the building, which is located at the center of the building's plan. The center of the resistance to these lateral forces is called the center of rigidity. If the building is symmetrical with respect to resistance, that is, if opposite sides of the building are equally rigid, then the center of rigidity will coincide with the center of gravity (plans B, C, and D), and torsional forces will be small. If the building is not symmetrical with respect to resistance, that is, if two opposite sides have very different rigidities, then the center of resistance will be located near the stiffer wall, resulting in larger torsional forces. This is the case in plan A, where the upper shear wall is much stiffer than the lower moment-resisting frame.

18. B Torsional forces result when lateral forces act on a building that has a rigid diaphragm, usually of concrete or steel, and in which the center of resistance of the lateral load resisting system (the center of rigidity) is not located at the center of mass of the building (B is correct, D is incorrect). Torsion can occur in any shape building (A is incorrect) and with any type of lateral load resisting system (C is incorrect).

19. C Wind pressures used in UBC design take into account the high wind speeds of hurricanes (A is incorrect). Buildings or surface irregularities interfere with the wind, and thus decrease the wind pressures (B is incorrect). C is correct: wind creates both positive inward pressures and negative outward pressures on wall surfaces. D is incorrect, because the wind pressures at corners and discontinuities are higher than those on typical wall elements.

20. D All four are examples of plan irregularities. I, II, and IV are plans with reentrant corners, and III is an example of torsional irregularity caused by unsymmetrical rigidities. Regular and symmetrical structures have better seismic response characteristics than irregular structures. Therefore, irregular structures in earthquake-prone areas should be avoided if possible.

21. A The formulas for both wind and seismic forces include an importance factor (I), which depends on the occupancy or use of the building. In this case, both the original warehouse occupancy and the meeting hall occupancy have an importance factor I of 1.0. Therefore, no additional design lateral force is required. See Table 16-K on page 10. Certain essential or hazardous facilities have an importance factor greater than 1.0, which has the effect of requiring such facilities to be designed for increased wind or seismic forces.

22. C Walls must be anchored to all floors and roofs that provide lateral support for the wall. The anchorage for masonry or concrete walls must resist the design seismic or wind forces perpendicular to the wall, or a minimum force of 200 pounds per lineal foot of wall. Anchor A in the detail serves this function (correct answer C). Vertical load is transferred from the roof joists to the ledger, usually by joist hangers, and from the ledger to the wall by anchor bolts. Lateral load is transferred from the roof diaphragm to the ledger by boundary nails, and from the ledger to the wall by anchor bolts. Plywood panels may be connected together by being nailed to blocking at the panel edges.

INDEX